INTERCEPTOR FIGHTERS

INTERCEPTOR FIGHTERS

FOR THE ROYAL AIR FORCE 1935-45

Michael J.F. Bowyer

Patrick Stephens, Wellingborough

In memory of four brave warriors—
Rusty, Timmy, Rufus and Theo

First published in 1984

British Library Cataloguing in Publication Data

Bowyer, Michael J. F.
   Interceptor Fighters for the Royal Air
   Force 1935-1945
   1. Fighter planes—Great Britain—History.
   I. Title
   623.74'64'0941   UG1242.F5

ISBN 0-85059-726-9

Patrick Stephens Limited is part of the
Thorsons Publishing Group

Photoset in 10 on 11pt Times by MJL Typesetting, Hitchin, Herts. Printed in Great Britain on 115gsm Fineblade coated cartridge, and bound by William Clowes Limited, Beccles, Suffolk, for the publishers, Patrick Stephens Limited, Denington Estate, Wellingborough, Northants, NN8 2QD, England.

Contents

Introduction

Hurricane and Spitfire are names as precious as any within Britain's entire history. Their mention in wartime, their appearance, hurried the nation's pulse. All knew that these aircraft were largely responsible for their survival. But there was far more to the fighter programme than the production of two fighters in sufficient numbers — just — to stem the German conquest. Indeed, there were occasions when both came close to cancellation in favour of ideas which would have proved far less successful. This volume considers a wide range of fighter designs on offer to the Royal Air Force between the 1930s and late 1940s.

All had features in common. They were piston-engined, and their armament was fixed. All were conceived as fast defenders, short duration interceptors for island defence and rapid turn around, except a heavy cannon fighter. Jet fighters rest beyond the parameters of this volume. Compared with the Germans, the British made less effort to hasten the introduction of jet aircraft, prefering to await high reliability turbines and ample power — not to mention a thorough investigation of tactics governing their use. What is interesting where British fighter designs are concerned is the swing from air-cooled engines to easily streamlined liquid-cooled power plants, with an eventual return to the former always favoured by many senior officers.

Although designed for peak performance at medium altitudes, coupled to the fastest compatible rates of climb, British fighters gradually forsook their improved high level performance for ground attack capability, for which role the less vulnerable air-cooled engines won renewed popularity. Airframe strength was generally such that heavy ground attack weapons and loads could be safely carried, making the Typhoon interceptor into a fine rocket-firing aircraft and prolonging the operational life of the Hurricane.

Excluded from consideration within this volume are fighters with traversing guns and those intended for long-range operation and often used for intruder work. Absent, too, are very specialised fighters.

Sources of material contained in the book are numerous. Basically they revolve around the author's interest and experience of RAF aircraft, which commenced in the early 1930s. Recently it has become possible to weave personal experience and prolonged study into a considerable amount of government and Service material generally released. Within this volume may thus be found background details relating to policies which brought about some of the designs and ideas. To list every reference item involved would be an almost impossible task. Many official files which have been consulted are now held within Classes AIR 2, AIR 16, AIR 20, AVIA 15 and AVIA 18 open for public inspection at the Public Record Office. Further material exists in a wide assortment of files, many initially seeming likely to have little to offer. Within Class AIR 10 and in the RAF Museum may be found Air Publications relating to fighters and their armament. In the use of items from this material Crown Copyright is acknowledged.

Whilst reading the book and considering the graphs, one must remember that aircraft performance figures quoted are those achieved in very experienced hands and almost all during official trials wherein manufacturers'.claims were explored. Speeds quoted are True Air Speeds correlated to equate still air conditions, the latter similarly allowed for in the climb capability. It must be pointed out, too, that individual aircraft capabilities varied, that the addition of even a small item could alter performance to quite a marked degree. Thus, where possible, some indication of the aircraft's state has been included and unqualified figures excluded. A Typhoon loaded to 13,000 lb, for example, is hardly likely to perform as well as when 2,000 lb lighter.

My thanks are due to the Staff of the Air Historical Branch (MoD) for help at an early stage of the compilation of this volume. I am grateful to the considerable assistance given by the Staff of the Royal Air Force Museum in tracing relevant Air Publications and for the permission to use RAF Museum photographs. I am equally grateful to the Imperial War Museum for being able to reproduce photographs.

Very special thanks are due to D.C. Greenman, Esq, MRAeS, of British Aerospace, Filton, who has contributed very fine drawings of some projected Bristol aircraft of the 1930s. To L. Whitehouse, Esq, of Dowty Ltd, alias 'Boulton Paul', go my many thanks for some most interesting material relating to wartime fighters and illustrations of them. Of some projects undertaken in the 1930-50 period little trace remains, so it would be interesting to hear from any reader who may have some material relating to them.

Michael J.F. Bowyer
Cambridge, January 1984

Glossary

A&AEE Aeroplane and Armament
Experimental Establishment;
Boscombe Down, September 1939
et seq
AASF Advanced Air Striking Force
ACAST Assistant Chief of the Air
Staff (Training) (with responsibility
for aircraft development)
ADGB Air Defence of Great Britain
AFDU Air Fighting Development Unit
AI Airborne Interception (radar)
BAFO British Air Forces of
Occupation (Germany)
CAS Chief of the Air Staff
cg Centre of gravity
CRD Controller of Research and
Development
cs(u) Constant speed (propeller) (unit)
DCAS Deputy Chief of the Air Staff
DTD Deputy or Director/Directorate
of Technical Development
DGRD Director General Research and
Development
DOR Director of Operational
Requirements
EAS Estimated Air Speed
F.10/35, etc. Aircraft specification, 'F'
meaning fighter, the leading number
that of the specification within the
year and finally the last two digits of
the year in question
FB Fighter Bomber

FR Fighter Reconnaissance
FS Full Supercharger gear in use
Hurricane II/P2, etc Hurricane
MkII/2nd production specification
IAS (or ASI) Indicated Air Speed (or
reversal)
IE Initial Equipment (of squadrons)
IR Immediate Reserve (of squadrons)
LF Low-altitude Fighter
LR Long Range
MAP Ministry of Aircraft Production
MB 2,3,4,5 Martin Baker fighters
MS Maximum Supercharger gear in use
NPL National Physical Laboratory
NS Napier Sabre
octane Hydrocarbon of the paraffin
series; high octane fuels, having good
anti-knock properties, do not detonate
readily during the power stroke
OR Operational Requirement (list of
requirements preceding a specification)
RAE Royal Aircraft Establishment,
Farnborough
RG Rolls-Royce Griffon
RM Rolls-Royce Merlin
RP Rocket Projectile
rpg Rounds per gun
R/T Radio Telephone/Telephony
TAS True Air Speed
TI Trial Installation
UE Usual Equipment
VCAS Vice-Chief of the Air Staff

Part One:
The fast single-seaters

In the mid-1930s the RAF was a biplane force. Even then, the contrast between bombers and fighters was stark, as when 1 Squadron's Furies lined up by some 'Ginnys' (RAF Museum 5413-4).

Chapter 1

Times of change

A silver swarm, they came buzzing and bouncing. At the rear the fast Furies — in the Air Staff's opinion less outstanding than sometimes suggested. King George V saluted his biplanes, amongst his companions aware of plans to yield replacements revolutionary in armament, shape and speed; utterly different from anything on parade at this 1935 Jubilee Review. Emphasis remained upon single-seat patrol fighters differing little from those which battled over France in World War I. Standby for quick reaction was far away. Although possessing increased engine power, improved reliability, higher speed and enhanced capability at greater heights, it was the mid-1930s before major decisions led to British fighters having dramatically improved armament.

Arrival of the Fairey Fox light bomber in 1925 — top speed 156 mph — had jolted the RAF by outpacing its fighters. Response followed in April 1926 with Specification F.9/26 for a day and night Gamecock/Siskin fighter replacement featuring metal construction, an air-cooled engine and two Vickers .303 in machine-guns. Then came a call for the F.10/27 high altitude interceptor. The former attracted nine prototypes and eventually led to a long line of successful Hawker fighters; the latter, ultimately, to the Spitfire. Fortuitously, Sydney Camm, Hawker's Chief Designer, was ready to break with convention by choosing for his Heron fighter a simple, light, strong, metal airframe fabric covered. With the wood and metal Hornbill, flown in May 1926, he again deviated from the normal air-cooled engine by selecting a 698 hp Rolls-Royce liquid-cooled Condor. The fastest fighter yet for the RAF, the Hornbill — top speed 187 mph — handled none too well and its cockpit was cramped. Very evident was the liquid-cooled engine's need for adequate radiator and cooling systems — something to reckon with throughout the application of such engines to fighters.

To F.9/26 Camm devised a Hornbill replacement, the Hawfinch, and Bristol produced their private venture steel-structured Type 105 Bulldog (450 hp Bristol Jupiter VII) first flown on May 17 1927. This, competing against the Armstrong Whitworth Starling, the Boulton & Paul Partridge and Gloster Goldfinch, outshone them all. A Mk II Bulldog was ordered in November 1927, featuring a longer fuselage to aid spin recovery. It first flew on January 21 1928 such was the ease of the development then. Squadron trials showed the aircraft faster and easier to maintain than the Hawfinch, while its air-cooled radial engine was reckoned less vulnerable in combat than the Hawfinch's. In August 1928, 23 F.17/28 Bulldog IIs (Jupiter VII) were ordered and delivery took place between

May and October 1929, initially to No 3 and 17 Squadrons. Subsequent orders were for the F.11/29 Bulldog IIa (Jupiter VIIF giving 490 bhp at 8,000 ft), delivery of the first of 92 ordered in May 1930 commencing the following October. Bulldog IIa orders eventually totalled 254 aircraft, production continuing to May 1933. They provided principal equipment for ten squadrons and remained in front-line service until the spring of 1937.

Greater frontal area and associated drag, features of radial engines, could be slightly mitigated by fitting a streamlining Townend cowling ring of 1927 design. The effect upon the Bulldog was minimal. Its reliable 9-cylinder 28.7-litre Jupiter was Bristol's answer to the Rolls-Royce Eagle and Napier Lion. It passed its type test in 1921 to become parent of radials worldwide, its output gradually increasing to 610 hp at 4,800 ft.

Hawker's disappointment at not getting the F.9/26 contract was lessened by the development of another pace-maker. The Air Ministry in May 1926 revealed plans for a fast day bomber, Fox successor, the 12/26. Sydney Camm promptly responded with the Hart, reminiscent in layout of his fighters and of similar performance. Working with Rolls-Royce he selected their new F XI liquid-cooled vee engine, the single-cylinder casting of which gave it high power/weight ratio. Concentration was upon extracting maximum power from the engine's features. First flown in June 1928, the Hart reached 184 mph at 5,000 ft and had a service ceiling of 21,350 ft. These figures vouched well for the engine's fighter suitability and it became available just as Specification F.20/27, a revised F.10/27, called for a fast single-seat day interceptor with 'capability to overtake, in the shortest possible time, an enemy passing vertically overhead at 20,000 ft and 150 mph'. Fast climb from standby would replace the standing patrol. Of major importance would be the 'fighting view' and a high degree of manoeuvrability. In response Bristol offered a Bulldog variant powered by their new Mercury and

Progress was such that Bristol Bulldog IIs like K1653 served until only six years before the Spitfire IX (R.S. Allam).

Gloster revamped their SS 18 producing the SS 19A, *J9125*, which, Jupiter VIIF-powered, flew in August 1928 and achieved 190 mph.

Hawker also chose a Bristol Mercury II (480 hp at 13,000 ft) for their submission, the Hornet. Experience, however, showed the advantages of the liquid-cooled 490 hp Rolls-Royce F XI Kestrel with its reduced drag, so Hawker's F.20/27 had this relatively light 760 lb engine. Compared with present day fighters, the Hornet (*J9682*) was but a light plane, tare weight 2,338 lb; flying weight 3,148 lb. Handsomely exceeding specification requirements, it climbed (loaded) to 10,000 ft in 4.3 mins and 20,000 ft in 11.25 mins — and there held a climb rate of 950 ft-per-min. A speed of 190 mph at 20,000 ft had been asked for — the Hornet clocked 196 mph and was even faster at 10,000 ft where 214 mph was recorded. Take off in a slight breeze was accomplished in a mere 100 yds, landing 186 yds; touch down being at 68 mph. Armament consisted of two Vickers Mk II .303 in machine-guns fuselage-mounted. Fuel capacity was 50 gal — 5 gal less than required. Official appraisal was that it was easy to fly, manoeuvrable, had good slow speed qualities, provided a clear view for the pilot and performed well for its intended role of point defence. A contract followed for the RAF's first 200 mph fighter. Defence of Portsmouth, Dover and such specific targets became the prerogative of its derivatives, 21 Hawker 13/30s ordered in August 1930 as Fury 1s (Kestrel IIS giving 485 hp at 11,500 ft and 575-590 hp at 13,000 ft).

The Hornet's performance confirmed the Air Staff's belief that the Bulldog belonged to a passing era. A far more radical fighter would be needed, and in quantity, by the mid-1930s judging by rapidly improving bomber performance. Therefore formulation commenced in March 1930 of F.7/30, the most important fighter requirement ever, calling for a single-seat day and night interceptor. Rapid climb would be coupled to good manoeuvrability and high speed at high altitude. The pilot's view was of paramount importance. Stability needed to be good and the aircraft highly controllable over a wide speed range. Production ease would be linked to simple maintenance. On August 5 1930 the prescribed armament was doubled from two to four guns and a heated cockpit asked for in a new F.7/30 draft approved by DTD on October 9 1930. Such was the economic situation though, that the specification had to be held in abeyance, from November 30 1930 to May 31 1931. On May 5 a new draft was compiled and the aircraft would now need to attain 15,000 ft in eight and a half mins and maintain at least 200 mph at 15,000 ft.

Despite the recession, Hawker had retained their Fury 1 contract, flying the first production example, *K1926,* on March 25 1931. By May sufficient Furies were at Tangmere for No 43 Squadron to equip. Their version, Kestrel IIS-powered, had an all-up weight of 3,317 lb, carried 50 gal of fuel and was armed with two Vickers II or III machine-guns. Its top speed, after rapid climb, was 205.5 mph at 15,000 ft. Service ceiling was little short of 30,000 ft; brakeless landing about 250 yds. Not until seven years later did a replacement type enter service.

Opposite
Top J9123, *the Hawker F20/27 prototype, upon which the Hawker Hornet and Fury were based* (RAF Museum P6870). **Centre** J9682, *the Hawker Hornet, immediate predecessor of the Fury.* **Bottom** *First flown on January 28 1928, the Bristol Type 105A Bulldog II prototype,* J9480.

The most famous Hawker Fury Is were those of No 1 Squadron. K5673, K2043, K2039 *and* K2881 *are illustrated* (RAF Museum 5717-2).

On October 9 1931 approval was given for the release of F.7/30 to industry, after repeated detail revisions. But the basic requirement remained, for a four-gun single-seat day and night fighter with a top speed of not less than 200 mph at 15,000 ft. Suggestions for a speed of 250 mph were not proceeded with. In order of importance the specification listed speed, rate of climb, manoeuvrability, ease of production and four guns — two in the wings and two in the fuselage. Hawker, eager to develop the Fury, flew, on April 13 1932, a private venture Intermediate Fury meeting ideas embodied in F.7/30. There were, though, far more attractive contenders to discuss when the Design Tender Conference opened on May 27 1932. It lasted a week, with over a dozen designs to consider. Some firms submitted several schemes — including biplanes, monoplanes, alternative engines, a three-engined pusher layout and twin-engined types. Prototype prices ranged between £4,500 and £9,000, proposed building periods extending between eight and 11 months.

Both Hawker designs were rejected. The monoplane was unexciting, the biplane obsolescent. Hawker then made it clear that they would proceed with a PV monoplane. Gloster tendered the SS 19, a further edition of *J9125*. Three designs were eventually selected for competitive trials, two of them biplanes from Blackburn and Westland. Attracting most attention, though, was the monoplane design from Supermarine, a company new to the fighter scene and applying expertise won through its Schneider Trophy float-planes and close association with Rolls-Royce.

Proposals submitted on February 20 1932 for their Type 224 were rewarded with a contract on August 2, for a prototype to be delivered in June 1933. This all-metal, gull-winged monoplane would be powered by a steam-cooled Rolls-Royce Goshawk II. A top speed of 238 mph at 15,000 ft was forecast — a good

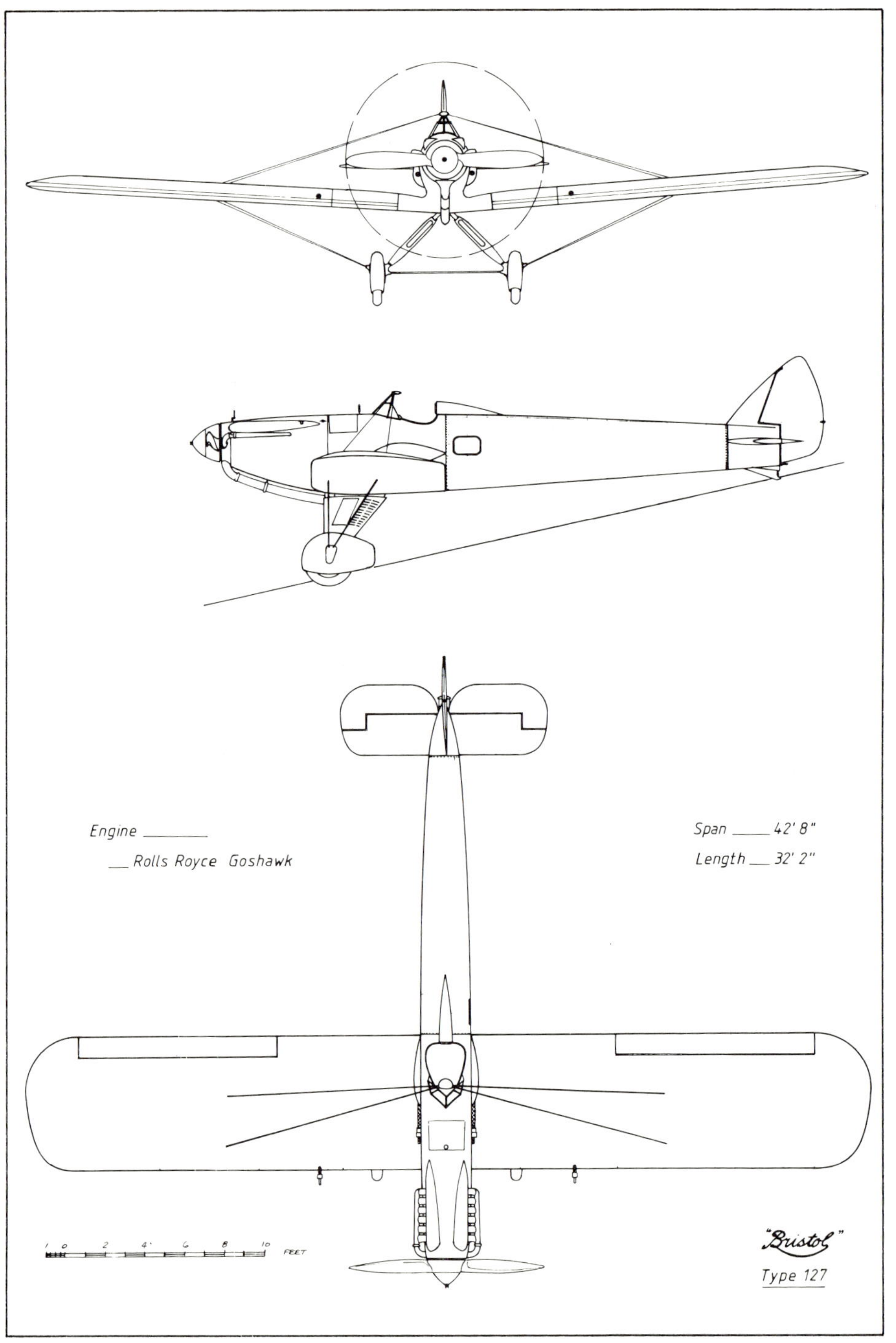

The Goshawk-powered Bristol Type 127, contender to Specification F.7/30 (D.C. Greenman).

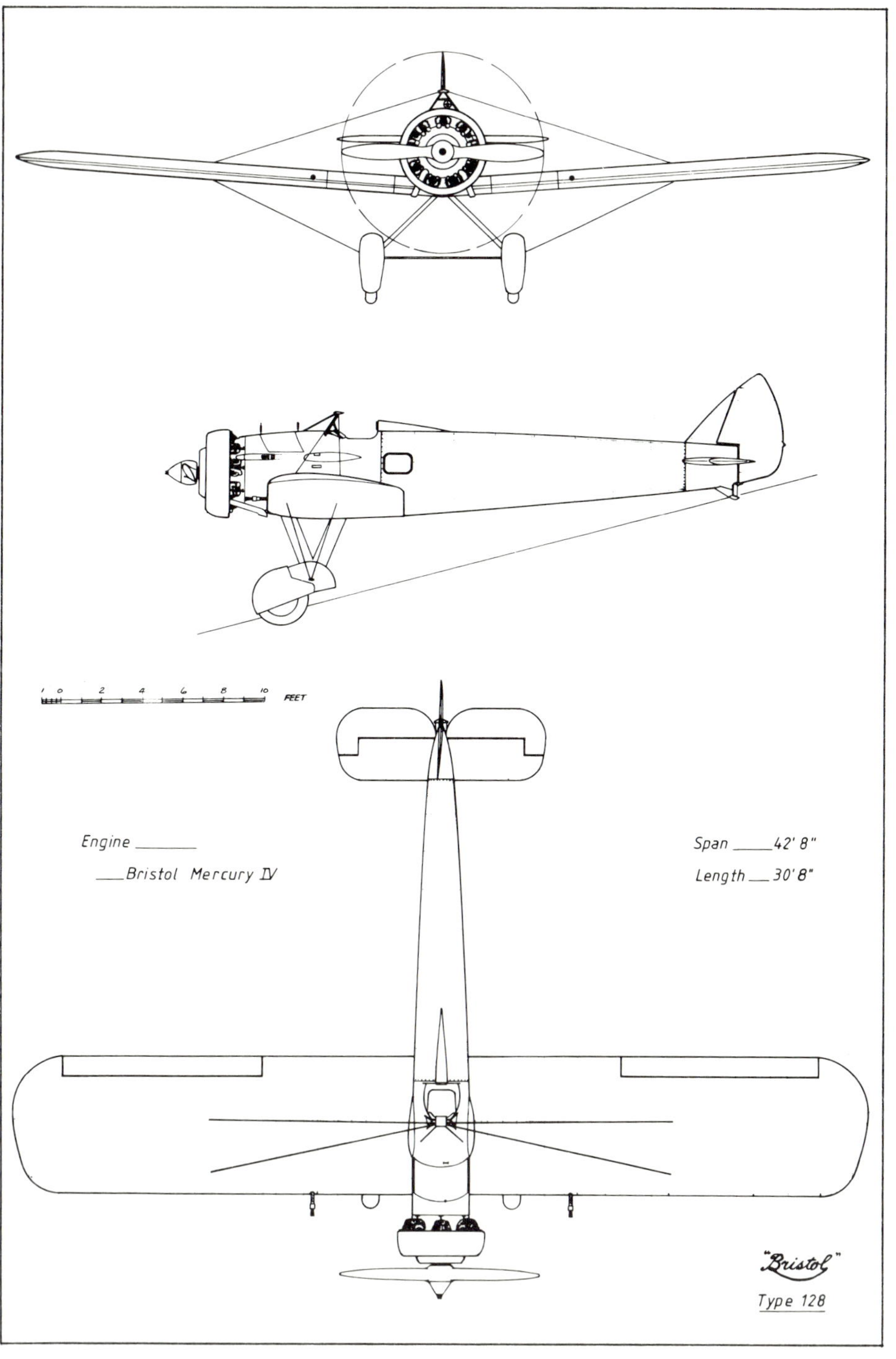

The Bristol Type 128 was similar to the Type 127, but would have been powered by a Bristol Mercury IV engine (D.C. Greenman).

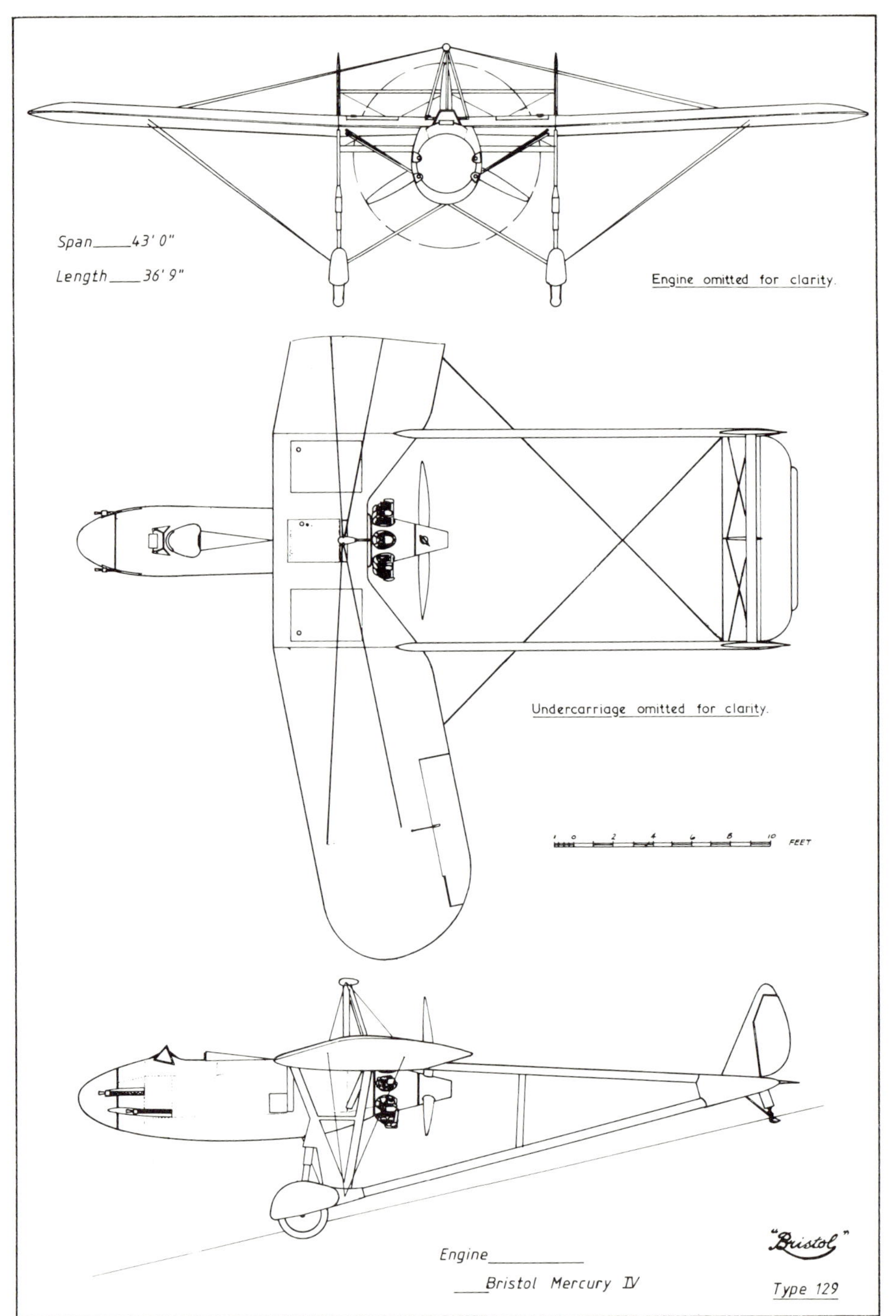

A most unusual-looking design to Specification F.7/30 was the Bristol Type 129 (D.C. Greenman).

advance on the Bulldog's 174 mph at 10,000 ft. Already it was plain that the Hawker Fury would soon be out-fought and on December 19 1932 at an Air Staff meeting a specialised fighter was suggested to replace it. Four days later it stood at fourth priority on the aircraft programme. Then the possibility of the Supermarine 224 fulfilling such a position was mooted. The Rolls-Royce 'R' engine of the Schneider racers was too heavy for a fighter and the Kestrel developed insufficient power. Supermarine, with little choice for their Type 224, opted for a Kestrel variant called the Goshawk which, being fed with 84 octane petrol, would give 690 hp at 3,000 ft in an aircraft carrying 80 gal of fuel and up to six guns with 2,000 RPG. Already its wing area had been increased to hold the landing speed at 66 mph.

A comparison was then undertaken between the Fury 1 and three feasible replacements:

	Fury 1	Fury Replacement	Supermarine 224	Revised Supermarine F.7/30
Loaded weight, lb	3,311	4,150	3,175	4,140
Wing loading, lb sq ft	13.2	16.5	15.05	14
Maximum speed at 15,000 ft, mph	205.5	261	244	238
Time to 20,000 ft, mins	9.4	8.23	9.0	10
Service ceiling, ft	30,000	36,500	36,200	36,000
Landing speed, mph	62.5	66	65	60

Fuel of higher than 77 octane was already judged essential for better performance, but the Treasury expressed concern even at the cost of an increase to 80.

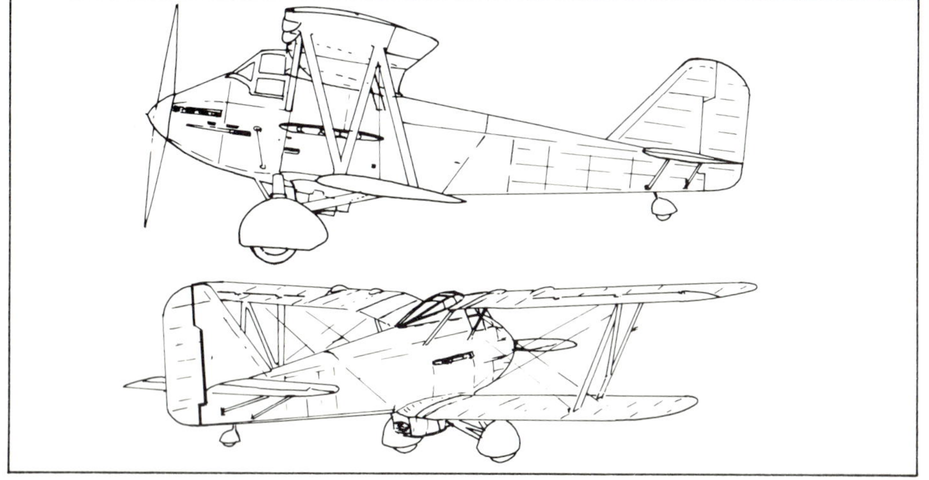

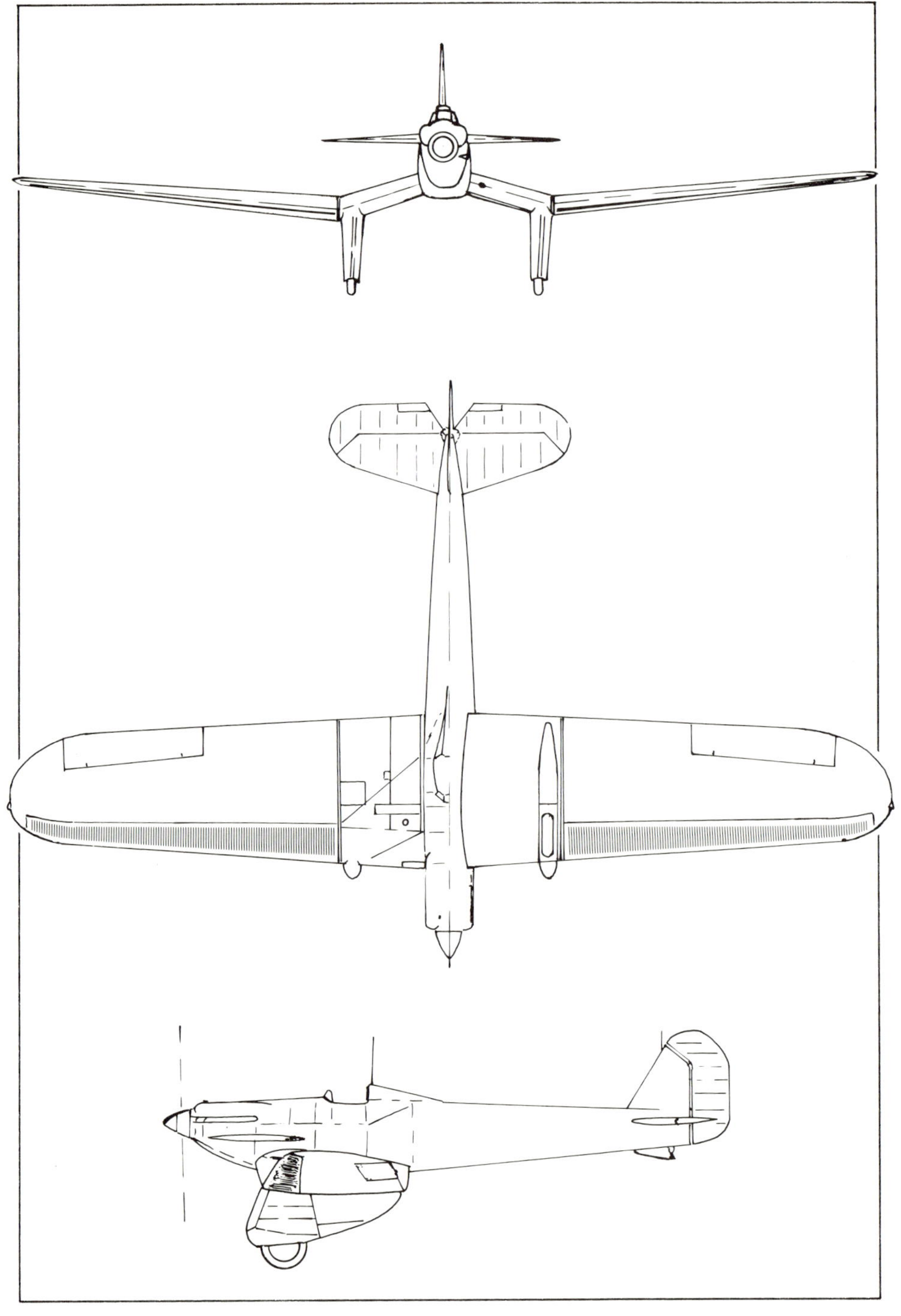

Left *The Westland F.7/30 biplane fighter.* **Above** *The Supermarine Type 224 to Specification F.7/30.*

Existing engines were unable to accept 84 octane fuel without major modifications, but advanced engine designs would suffer unless the octane level was raised. Uprating must come about.

By March 1933 Supermarine's F.7/30 was the most favoured future home defence interceptor. Now incorporated was full night operating capability — reducing the performance.

Good as it had seemed in 1931, the Fury was judged 'a failure, having insufficient margin of speed over aircraft such as the Hart to ensure interception of raiders', according to the Air Staff. Furies would now operate back in the anti-aircraft zone in wartime. Yet clearly, any day and night fighter would be at a disadvantage because bomber design was 'improving to such an extent that, even with the highest performance day fighter, the chance of interception upon the inward journey (was) remote'. Compromise fighters would be of little value for high-speed combat, although combining plans for a two-seater with the day/night concept seemed reasonable. To achieve necessary high performance the Air Staff remained adamant that higher octane fuel, with its greater compression ratio for increased power output, was essential. They soon had their way for in 1933 DTD230 87 octane, the first leaded fuel, was introduced. This was followed, experimentally, by 100 octane petrol in 1934.

Deliberations of March 18 1933 became linked to an event of major importance. The Air Staff had before them plans for the Supermarine F.7/30 'Bulldog Replacement day/night fighter' as well as the four-gun Fury specialised fighter replacement now being planned by Hawker to carry six guns. What seemed incredible was a suggestion now that the latter might even carry eight machine-guns. Comparative figures for the four projects showed:

	4-gun F.7/30	Fury replacement		
		4-gun	6-gun	8-gun
Maximum speed at 15,000 ft, mph	240	265	261	256
Time to 20,000 ft, mins	8.2	7.8	8.2	8.6
Service ceiling, ft	38,000	37,500	36,500	35,400
Landing speed, mph	60	66	66	66
Normal endurance, hrs	2.5	1.7	1.7	1.7

All were fast, feasible, heavily armed fighters rendering the 'Bulldog Replacement' as originally conceived a non-starter. Quite clearly a 'Fury Replacement', including all the F.7/30 ideas, was a practicable proposition. If a suitable engine using high octane fuel was devised then performance could be far better than forecast.

The Supermarine 224 mock-up was already three months behind due to alterations and a conference planned for April 20 1933 was postponed amid official concern. Hawker, intending not to be left out, had already fitted a 604 bhp Kestrel VIS in their High Speed Fury *K3586* which first flew in May 1933. For the next three years its only value was as an engine test bed, although it roughly equated F.14/32's needs.

When the Supermarine 224 mock-up conference eventually took place in July 1933 the programme was lagging badly. Linked to a six month contract extension was a warning that, if the prototype did not reach Martlesham by

December 31 1933, it was unlikely that the aircraft would be adopted for service.

Meanwhile, other manufacturers vying for the lucrative order were constructing an astonishing assortment of F.7/30s. Bristol devised a conventional Bulldog development whose Mercury VIS gave it a creditable speed of 224 mph. Blackburn's F 3 *(K2892)* built to Ministry order was of most unusual layout. Metal wings were fabric covered, whereas the metal fuselage featured Dural skinning. Positioned between the fuselage and the lower mainplane was the radiator. Riding high was the pilot, given splendid view. Engine cooling and structural problems afflicted the prototype when it appeared in July 1934, by which time its predicted top speed of 190 mph rendered it of no value. Blackburn never bothered to fly it!

Westland originally received a contract for a monoplane, also of unusual layout since the propeller was joined to the centrally sited engine by a long shaft over which the cockpit was situated. Fears that the landing speed would be too high led to its conversion into a biplane, the upper mainplane being located immediately aft of the cockpit. Martlesham tests at a flying weight of 5,207 lb showed a top speed of 146.5 mph at 10,000 ft reached in 8 mins and 122 mph at 20,000 ft after an 18.8-min climb. Thus, its performance fell short of requirements.

Bristol made two further attempts to acquire the F.7/30 contract, initially with their Type 123 first flown on June 12 1934. This was a squat biplane which met with severe stability problems. Secondly came their Type 133 radial-engined inverted gull wing monoplane powered by a Mercury VIS 2 and first flown on June 8 1934. Although a better performer than others, it crashed prior to official trials. Three other Bristol projects, Types 127, 128 Goshawk/Mercury monoplanes and the Type 129 pusher monoplane, were abandoned.

With Air Staff discussion in August 1933 again ranging around a combined F.7/30/Fury replacement, Major Buchanan, DDTD, decided to visit Hawker to hear first-hand of their private venture 'Fury Monoplane' upon which they had started work after failing to secure an F.7/30 contract. Planned around a vapour-cooled Goshawk, it had four machine-guns, two wing mounted. Buchanan was impressed with what he heard and saw, but this machine was very much for the long term. With all the new fighters so far off and the need to replace the Bulldog looming fast, an interim order was placed for 24 two-gun Gloster SS 19B Gauntlet biplanes derived from the company's work with *J9125* over a long period. Gloster was delighted for again they were in the fighter stakes, and would remain so until the 1960s. Sufficient Gauntlets were being acquired only to replace No 19 Squadron's Bulldogs. Could the company obtain more orders?

A long haul for Gloster had at least been rewarded. They had tendered their Goldfinch to F.9/26, but its fuel load was insufficient. H.P. Folland, their designer, then devised a Mercury IIA variant overtaken by F.20/27's requirements. Gloster went ahead privately with the SS 18 *(J9125)*, a metal fabric-covered biplane — strong and two-bay — which flew in January 1929. Although it handled well its top speed was only 183 mph at 10,000 ft and it took 13.5 mins to reach 20,000 ft because its Mercury engined failed to provide the promised 500 hp. Folland redesigned it to take a 480 hp Jupiter VIIF, the aircraft becoming the SS 18A in mid-1929. It was still under-powered so a 560 hp Armstrong Siddeley Panther III two-row radial was tried. Although the engine was

Above *The six-gun Gloster SS 19,* J9125 (RAF Museum P8840). **Below** J9125 *as the Gloster SS 19B closely resembled the Gauntlet I. Spats have been removed to avoid mud clogging)* (RAF Museum P6798).

very heavy it raised the top speed to 205 mph at 10,000 ft.

With further refinement and the Jupiter VIIF re-instated, the SS 19 arrived, flying in summer 1930 with two Vickers guns in the fuselage and four Lewis guns in the wings. Two were attached below the top mainplanes, two beneath the lower, all reliable but each carrying only one drum of ammunition. This version had a top speed of 188 mph at 10,000 ft, weighed 3,520 lb loaded and could

The framework of a typical fighter of the 1930s, in this case a Gloster Gauntlet I.

attain 20,000 ft in about 15 mins. It still did not meet official needs. Wheel spats were added and further refinements before, as the SS 19A, it was submitted for official trials in November 1931. With a speed of 204 mph at 10,000 ft it proved much faster than the standard Bulldog IIa. Installation of a reliable 536 hp Mercury VIS raised the SS 19B's speed to 212 mph at 14,500 ft coupled to a climb to 20,000 ft in 12 mins 15 secs which, for a 3,858 lb single-seat fighter, was good. Upon this biplane the Gauntlet I was based.

Specification 24/33 was drawn up in September 1933 and the contract signed in February 1934 for 24 Gauntlet 1s to be built in about March 1935. The Mercury VIS provided 575-605 bhp at 12,500 ft and 615-645 bhp at 15,000 ft, at 2,400 and 2,750 revs respectively. Two Vickers guns were fuselage-mounted to avoid risk of jamming and the pilot being unable to clear them.

Slow progress with monoplane fighters had thus brought a new biplane into service, which was unintentional. In great secrecy, though, Hawker and Supermarine were busy. On December 5 1933 Hawker completed general arrangement diagrams of their Goshawk-powered Fury Monoplane, presenting them at the Air Ministry on December 18 1933. Discussion produced a suggestion that a superior aeroplane would evolve if powered by the new Rolls-Royce engine, the 1,000 hp PV-12. In January 1934 Hawker agreed to its incorporation in their project. A scaled-up version of the Rolls-Royce Kestrel (whose output was to rise to 745 hp by 1938), the prototype PV-12 was completed in 1933. Its cylinder block had a crankcase top half as a one-piece casting, and its revised cooling system was based upon the Goshawk's. Early tests led to ethylene glycol being used as a coolant in this, the engine which became the Rolls-Royce Merlin.

December 1933 came and still the Supermarine 224 had not been delivered. Delay was being caused by its complicated power plant and Supermarine were trying hard to reduce radiator weight. A larger wing had been fitted to cut wing loading for the aircraft's weight had risen to 4,000 lb and, hopefully, its speed to

240 mph. Non-delivery of equipment, the company maintained, largely delayed their prototype. Rumour — never doubted by the Air Ministry — was that Supermarine's talented designer, R.J. Mitchell, was privately concentrating upon a far superior invention.

On February 20 1934 the Type 224 *(K2890)* eventually flew, an unusual looking machine with an inverted gull wing and hefty, trousered undercarriage. It featured a monocoque fuselage and stressed metal skin. Instead of 240 mph its top speed was 228 mph and it took 9.5 mins to reach 15,000 ft — not the forecast 6.6 mins. Problems with the condenser system in the wing leading edge made it obvious that the Type 224 could never evolve into a front line fighter.

Gloster, very pleased with the recent contract and upon hearing of the 800 hp Mercury ME3S, modified an SS 19-type fuselage to take this powerful motor. Hawker had put an improved Kestrel VI in Fury *K1935,* trying again for an F.14/32 order, and found their latest fighter could reach 228 mph. Their main effort, though, had switched to the 'Interceptor Fury Monoplane', first drawings of which reached their Experimental Section in May 1934.

Rolls-Royce and Bristol were not the only firms offering new power plants, Air Staff interest being aroused in the Fairey Prince 24-cylinder 'double' engine. Each of its two banks of 12 cylinders drove one of two three-bladed co-axial propellers independent of others. Powerful, but complex.

By May 1934 a pure Fury replacement with rapid climb to 20,000 ft was no longer needed. Instead, there must be the fastest possible fighter having the best possible climb rate to 20,000 ft. Four compromise fighter schemes were under consideration, all Prince-powered. Although four guns meant lower wing loading and improved climb, such an aircraft's higher speed would increase its landing run. Four permutations were under review:

	8-gun	6-gun	4-gun 'x'	4-gun 'y'
Flying weight, lb	5,020	4,800	4,500	4,580
Maximum speed at 15,000 ft, mph	262	263	264	256
Time to 20,000 ft, mins	9.6	9	8.4	7.7
Service ceiling, ft	32,000	33,000	34,000	35,000
Landing speed, mph	70	70	70	60
Normal endurance, hrs	1.6	1.67	1.67	1.67

All would have brief endurance, an unfortunate future feature of so many British single-seat fighters, and soon it was obvious that the Prince was an engine for the far future. In June Hawker submitted a 1/10 scale model of their 'Fury Monoplane' to the NPL for aerodynamic tests. Sydney Camm hoped the prototype — forecast weight 4,600 lb — would fly in spring 1935.

June 1934 found the Air Staff convinced that the future lay largely with fighters less specialised than the Fury, speedy aircraft able to climb fast to deliver a rapid, accurate, hefty punch. Details of the secret 20 mm Hispano cannon were difficult to obtain from France and, although effective, the first .50 in machine-guns were heavy and slow-firing. Since 1918 the RAF had made much use of Vickers guns, but these were outdated. Investing in a new gun, however, made sense but would be costly. Fighter armament had already doubled from two to four guns and by doubling that again the lethality within a brief time would be great. Investigation showed that eight Browning guns would be ideal. Deliberations of 1933 and 1934 were then drawn together to form the basis of an eight-gun fighter to replace the Fury and all derivatives of F.7/30. A preamble to the detailed specification explained that 'speed excess of modern fighters over contemporary bombers has so reduced the chance of repeated attacks by the same fighters that it has become essential to obtain decisive results in the short space of time offered by one pass only. The object is to produce a day fighter in which speed in overtaking the enemy at 15,000 feet, and rapid climb to height, are of primary importance. The best speed possible must be provided at heights between 500 feet and 15,000 feet. In conjunction with performance, maximum hitting power must be provided. For this purpose it is proposed to mount eight machine-guns. In order to assist performance, by reducing body size, none of these need be mounted in the fuselage within reach of the pilot.'

Maximum speed was to be not less than 275 mph, with climb to 20,000 ft a secondary aspect but the best possible. Service ceiling needed to be at least 33,000 ft, endurance 1.25 hrs at normal revs at 20,000 ft, allowance being made for ½ hr flying at full throttle. Landing run must not exceed 600 yds. Eight mechanically- or electrically-fired machine-guns, probably Colt Brownings, would fire 734 rounds in 5 secs, total load 1,100 rounds. A downward field of view 10° over the nose was required, along with good all-round view for search and formation flying, an advantage the monoplane offered. Good manoevrability was needed and the enclosed cockpit would feature R/T, oxygen supply and heating. The undercarriage would be retractable.

Approval of these ideas was not universal, some senior officers arguing for traversing guns. Many squadron pilots maintained preference for guns close to hand in case of stoppages. There was concern over flying control at slow speeds, whilst it was pointed out that if it took five years for a fighter to evolve — as seemed the pattern — then this new one must reach a speed above 300 mph, even 350 mph. Improvement in radio control of interception evident in the 1934 air exercises emphasised that climb to intercept was no longer as important as speed. Commander-in-Chief of the ADGB considered that good flying characteristics to around 25,000 ft were more important than higher ceiling. One suggestion was for the retractable undercarriage to be sufficiently rigid to be lowered for use as an air brake, improving control in combat. All of this brought a detailed review of the merits of British fighters compared with foreign equivalents following R.J. Mitchell's submission, on July 27, of plans for a radically revised F.7/30. This featured thin, straight wings and tailplane, small area flaps, retractable undercarriage and oil tank with cooler beneath the 1,000 hp PV-12 engine. Mitchell reckoned that he could have the aircraft flying in early 1935, forecasting its speed as 265 mph — higher if the PV-12 improved. The Director of Technical Development replied that in June it had been decided that all future day fighters would carry eight wing guns — this new project unacceptably had four.

Sydney Camm's ideas were also complete and Hawker ready to build a prototype. On September 4 1934 his plans reached the Air Ministry, the company now hoping for government support and knowing that the Air Staff need now was for a fighter reaching 310 mph at 15,000 ft. Between July 27 and August 23 discussion revolved around applying eight guns to both these advanced designs. Squadron Leader R.S. Sorley pointed out that twin-engined bombers were equalling the speeds of the fastest fighters, making the chances of repeat passes slim. Conclusive, 2-sec engagements with heavy weight of fire were therefore essential. In trial firings eight Browning guns had recently fired 256 rounds-per-sec and manufacturers wanted to fix eight guns in their fighters. The problem was how to mount them without thicker wings. Wing-mounting demanded rigid structure and complicated firing systems. Thicker wings of the Hawker design allowed for simpler installation but the prototype was too advanced for modification without causing undue delay.

Although the case for an eight-gun firepower was unassailable, prospects of wing guns jamming demanded high reliability. Speed remained the vital requirement, along with easy maintenance and rapid re-arming, not to mention ease of production. Future demands were rapidly outstripping each specification, so it was decided to invite tenders for the specialised F.5/34 Fury Replacement, then quickly formulate Specification F.10/35 for a fighter attaining 310 mph at 15,000 ft and combining features of both new monoplanes. None could emerge quickly irrespective of urgency. Therefore the Air Staff was relieved to find Gloster producing another fast interim biplane. On September 12 1934 the company flew their SS 37 private venture four-gun F.7/30 design featuring single bay, stronger spar wings than the Gauntlet and no axle spreaderbar between the Dowty internally sprung wheels. Mercury VIS-powered, it reached a creditable 242 mph at 15,000 ft whence it had climbed in 6.5 mins. At 30,000 ft, reached in 21.6 mins, its top speed was 210 mph after a take off weight of 4,390 lb. Gloster hoped to later fit an 800 hp Mercury ME3S, giving at least 250 mph, or a suggested Bristol Perseus.

Tenders for F.5/34 were invited from Bristol, Gloster and Westland on November 16 1934. Meanwhile, attempts were being made to persuade Supermarine to fit a 24-cylinder 725 hp Napier Dagger in their latest design to relieve pressure on Rolls-Royce production lines. Supermarine strongly resisted the notion before they received an order on December 1 1934 for one prototype Supermarine 300, for delivery ten months hence. Firm agreement that it would have neither a Goshawk nor a Dagger, but a PV-12, was reached on December 6 after a forecast of this engine giving 1,025 hp at 15,000 ft on 2,800 revs and having a dry weight of 1,200 lb. The aircraft's Specification, 37/34, was sent to Supermarine on December 22 1934. Hawker received their prize as a New Year offering bringing the Spitfire and Hurricane into reality. Dramatic indeed were the changes afoot, and two of the most famous of aeroplanes would soon be in the headlines.

Chapter 2

The first for the few

Royal Air Force expansion was well underway by 1935. Firms were asked for their F.5/34 tenders by February, the Hawker Fury Monoplane mock-up was officially viewed on January 10 and five days later a fuller specification for F.37/34 arrived at Supermarine. The first Gauntlet 1 had flown in December 1934, its operational weight being 3,937 lb. Service trials showed a top speed of 215.5 mph at 16,500 ft with 20,000 ft being reached in 11.43 mins. The service ceiling was 33,200 ft. This good performance was attributed to thin wings, careful wing bracing and attention to detail. On February 18 1935 No 19 Squadron received its first Gauntlet I (*K4086*) for trials. Bulldog replacement had started.

Impressed with the Hawker Monoplane, the Air Ministry ordered a prototype on February 21 1935 — to the design of the previous September — as the F.36/34. Estimated all-up weight was now 4,900 lb and top speed 330 mph at 15,000 ft. To help Hawker, the company received in March 1935 a production contract — largely sub-contracted — for the 6/35 Hawker Fury II Kestrel VI-powered and carrying additional fuel. Gloster, taken over by Hawker in February 1934, received an order for 104 Gauntlet IIs featuring Hawker-type construction which made them easier to produce and maintain.

Gloster Gauntlet I K4095 served with 19 Squadron from May 1935 until December 1938, then saw Middle East service.

The prototype Hawker Fury II, K1935, was built in 1931. A modified Mk I, it was fitted with a Kestrel VI and had a top speed of 228 mph (RAF Museum P6882).

Air Staff viewing of the impressive Supermarine F.37/34 mock-up took place on April 26 1935. They revealed to R.J. Mitchell ideas for a 1940's fighter, the 310 mph F.10/35. But when the likely performance of the F.37/34 was scrutinised it was clear that this might equate the planned F.10/35. Four more guns could be carried if the wing chord was increased. On May 1 1935 Squadron Leader R.S. Sorley minuted the Assistant Chief of the Air Staff that modifications bringing the two fighters into line with F.10/35 mainly concerned armament. Whereas the Hawker could have eight guns set in two blocks of four, easy to harmonise, the Supermarine would need them well spaced, spreading the load throughout its thin wings. A loss, not viewed with much concern at the time, was a fuel reduction of from 95 to 75 gal. The Spitfire's resulting thin, elliptical, broad-chord wing owed plenty to logic and expediency. The DOR's Staff, of the opinion that both monoplanes could enter service in 1936, suggested ordering off the drawing board and cancellation of the Gauntlet II. The Air Staff would not risk that, rightly sensing delays. Mitchell claimed that his F.37/34 would start official trials in October 1935, but it did not fly until March 1936.

June 1935 brought an enquiry directed at contractors interested in F.5/34 as to whether any of their designs could equate F.10/35. The F.5/34 was a day fighter and adding night capability could increase its fuel load, wing area, etc, and generally inhibit it.

Meanwhile Gloster was extracting everything possible from the SS 37, proposing a Mercury IX version and an enclosed cockpit. Impressed with the prototype, the Air Ministry ordered 23 as Gladiators against Specification F.14/35 in case of more monoplane delays. In modified form *K5200* reached Martlesham in July 1935, by which time 19 Squadron's Gauntlets had made a profoundly good impression.

On July 20 1935 the F.36/34 contract was amended to call for eight wing guns

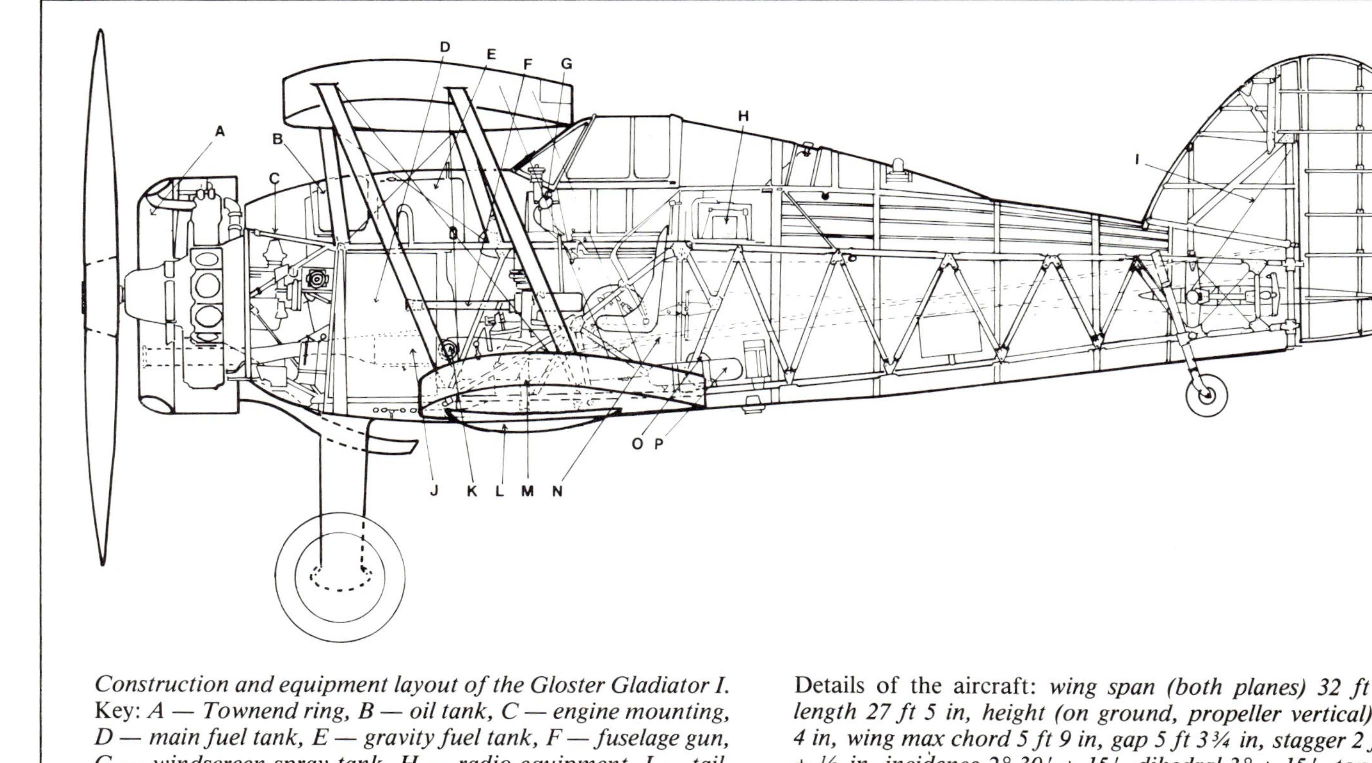

Construction and equipment layout of the Gloster Gladiator I. Key: A — Townend ring, B — oil tank, C — engine mounting, D — main fuel tank, E — gravity fuel tank, F — fuselage gun, G — windscreen spray tank, H — radio equipment, I — tail-bracing wire, J — Gallay oil cooler, K — cockpit heater, L — under-wing blister over gun, M — ammunition box, N — control wires, O — compressed air cylinder, P — oxygen cylinder.

Details of the aircraft: *wing span (both planes) 32 ft 3 in, length 27 ft 5 in, height (on ground, propeller vertical) 10 ft 4 in, wing max chord 5 ft 9 in, gap 5 ft 3¾ in, stagger 2 ft 3 in ± ⅛ in, incidence 2° 30' ± 15', dihedral 3° ± 15', top plane area 169.2 sq ft, bottom plane area 153.8 sq ft, tailplane area 19.3 sq ft, elevator area 18.24 sq ft, track (wheels off ground) 7 ft 2.4 in, propeller disc diameter 10 ft 9 in (2-bladed), main tank holds 63 gal, gravity tank 20 gal.*

(Vickers Mk V or Browning), ten days after Hawker commenced work on stressed metal-skinned outer wings. Firing trials commenced of the eight-gun wing on August 23 1935 with the certainty that the aircraft could not be delivered in 1936, perhaps not even until 1937. Therefore, in September 1935, another 100 Gauntlet IIs were ordered.

On October 23 1935 Gladiator prototype *K5200*, fitted with a Fairey Reed three-bladed propeller for comparative trials with the wooden Watts propeller and for interrupter gear tests, reached Martlesham. Three-bladed propellers were still rare but increased engine power demanded them. It was discovered that the Gladiator's speed rose to 253 mph at 14,500 ft although its weight increased to 4,456 lb and take off run to 450 yds.

As the Gladiator was flying to Martlesham the Hawker F.36/34 was being transported by road from Kingston-on-Thames to Brooklands, for flight trials. Weighing on October 30 revealed a flying weight of 5,416 lb — a considerable increase over that expected. Nevertheless, when P.G. Bullman took *K5083* on its first flight, on November 6, he reported on its excellence. On February 7 1936 it made its way to Martlesham for intial appraisal. Installed was Rolls-Royce Merlin 'C' No 15, driving a two-blade Watts propeller. Oil temperatures had run high, necessitating this replacement engine. Martlesham's pilots were impressed — 'simple and easy to fly . . . no apparent vices . . . all controls operate satisfactorily and are excellent at low speeds . . . aileron and rudder controls became too heavy at high speeds . . . take off and landing present no difficulties and in spite of the high top speed there is not difficulty in making an approach on a normal aerodrome when the undercarriage and flaps are down.'

Flying weight on trials was 5,672 lb, including 107.5 gal of petrol and 8 gal of oil. Highest recorded speed was 315 mph at 16,200 ft, reached in about 6 mins. Climb to 30,000 ft took 18.1 mins and fell away from about 2,300 ft per min after 13,000 ft. Unstick in a 5 mph wind with radiator flap shut took 265 yds; to climb over 50 ft a 430 yd run — less than the Gladiator's. Landing over a 50 ft screen was made in 475 yds, speed being 70 mph.

Whilst the Hawker machine was being scrutinised at Martlesham, the prototype Spitfire *K5084* undertook, on March 5 1936, its maiden flight from Eastleigh. Supermarine had constructed their prototype in little over a year. Since it embodied many new techniques, they had done well. Within a few weeks both new fighters had proven their excellence, production orders being awarded on June 3 1936. Expansion Scheme F required 900 new single-seat fighters to be in service by March 31 1939. Of these, 600 were to be F.36/34s, the remainder from Supermarine. Hawker — reckoned more ready to cope with heavy production — would be concentrating upon a type seemingly easier to build and probably more robust. In addition, it was ahead for, three months previously, Hawker had planned its production. It was, however, produced with less development than the advanced Spitfire.

Allowing expansion of the fighter force, Gauntlet II deliveries beginning on March 27 1936 were welcome. RAE Farnborough tried a canopy on a Gauntlet, but the engine threw back too much oil for this to be practicable although the problem was overcome in time for the Gladiator to be so fitted. Gauntlet IIs were cleared to have a new governor and three-bladed propeller and had a radio aft as a permanent balance fitting.

Acceptance trials of the Hawker Monoplane proved most satisfactory and on June 27 1936 the name 'Hurricane' was bestowed upon it. Supermarine's fighter

Top *Hurricane prototype K5083 early in its career. Fairing doors are attached to the undercarriage covers. Cockpit canopy modifications followed.* **Above** *Hurricane K5083, now fitted with an aerial mast. Note the retracted tailwheel* (Rolls-Royce WP6151). **Below** K5054, *the Spitfire prototype* (via Bruce Robertson).

became the legendary Spitfire on July 28 1936. A thoroughbred from the start, the Spitfire prototype arrived at Martlesham for tests on May 26 1936. Comments upon its trials were extremely favourable. Taking off at about 5,330 lb, its ailerons were light and effective, response being as quick as with the rudder and elevators. Cockpit layout seemed good and the aircraft was very stable in level flight. Aerobatics were accomplished with ease, take off and landing being normal with take-off swing easy to correct. The undercarriage could be raised in 10 secs and lowered in 15. Forward and rear view were fair, upward very good, but the curved windscreen caused severe distortion. Gun mountings also needed modifications. The aircraft was 'simple to fly and had no vices'. Recorded performance figures at loaded weight were: maximum speed 349 mph at 16,800 ft; maximum cruising speed 311 mph at 15,000 ft; time to 15,000 ft 5.7 mins; 17 mins to 30,000 ft; service ceiling 34,500 ft.

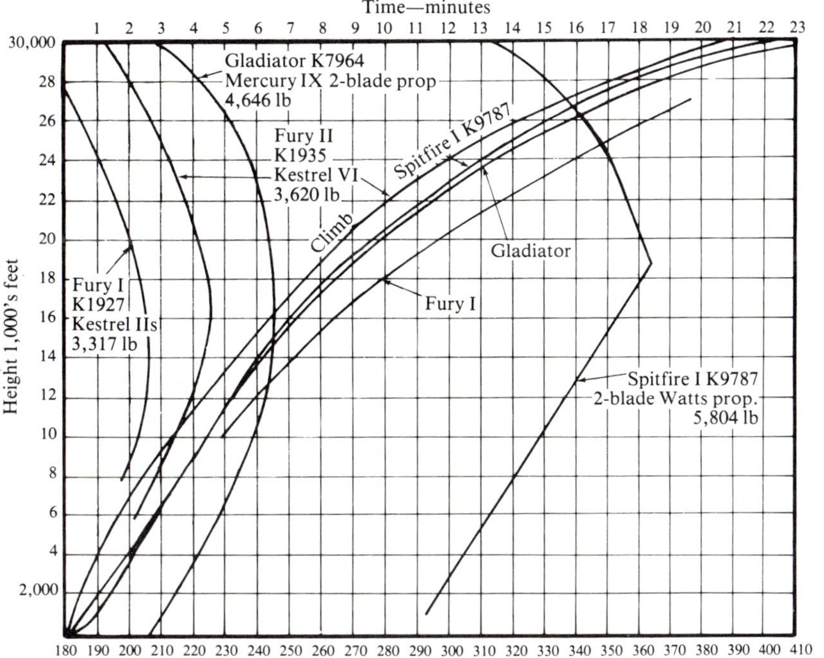

Comparison of top speed and climb of the Fury, Gladiator and Spitfire I.

Already the first F.5/34 contender was flying, the Vickers PV, commonly called the Venom. Vickers had earlier designed a radial-engined, low-wing monoplane fighter to F.20/27, the Type 151 Jockey with a Mercury IIA. First flown in April 1930, it suffered considerable vibration problems. A Townend ring was wrapped around the engine, and retained when a Jupiter VIIF was installed in January 1932. The rear fuselage, source of the vibration, was rebuilt, but the aircraft crashed in June 1932 following a flat spin. Vickers updated their Type 151 around a 625 hp Bristol Aquila sleeve valve engine. Eight machine-guns were installed — before the Venom's first flight on June 17 1936 — in thick, high lift section wings. A stressed, flush riveted, light alloy skin aided its performance. Like many radial-engined fighters it was very manoeuvrable and reached 312 mph at 16,250 ft during maker's trials. Both its all-up weight of 4,150 lb and wing area of 146 sq ft were much below the Spitfire's. Unlucky was the choice of an Aquila, development of which soon ceased. In any case Vickers would have the Spitfire to produce. What that company may not have known in 1936 was that the Air Staff had far from dismissed the Venom from an RAF career. An RAE pilot assessed it during the 1936 Hendon display and was impressed. Its 10 per cent propeller pitch variation was insufficient, 20 per cent, at least, being needed. The Aquila's big-end had limited life and material for engine sleeve valves had been changed from cast iron to cast steel, necessitating a nitrating process, which cast iron was intended to replace. Vickers, failing to get a contract, enquired whether their fighter might be exported. The Air Ministry replied that, since it featured a new specification, they could not agree to this, but would review it when the Hurricane and Spitfire were in service.

Actually the Air Staff were quite impressed by all they heard about the Venom and still favoured air-cooled engines for fighters. The Venom's performance was reckoned as good as the Hurricane's and it was said to be more manoeuvrable throughout its speed range. Easy to land, despite its high wing loading, it possessed a remarkably short landing run. Forward view was superior to that of the Spitfire and as good as the Hurricane's when taxying. But, 'it has appeared too late for inclusion in the present production programme', wrote the Deputy Chief of the Air Staff, Air Marshall C. L. Courtenay. The only hope for it might have come from special Treasury funding, or re-allocation of the ratio of bombers to fighters with more of the latter. Vickers, already stretched production-wise, would have needed others to build Venoms. Aquila production was not yet planned, Bristol being too pre-occupied with other engines.

The fighter programme currently envisaged 21 squadrons equipped with Hurricanes and Spitfires, delivery of which would commence in October 1937. Some 500 Hurricanes and 270 Spitfires were scheduled for delivery by March 1939. Possible extension orders for 1939-40 delivery, about 50 of each type, were under review. Introduction of a further fighter type would certainly have upset planning. Production would be of the Hurricane 1 (Merlin II) to

The Vickers Venom PVO-10 private venture fighter (Vickers).

Specification 15/36 agreed on July 20 1936 and the Spitfire 1 (Merlin II) to 16/36 agreed on July 28 1936.

A modified fighter re-equipment programme of mid-1935 called for 35 squadrons. Eight single-seat fighter squadrons would equip each of two Fighter Groups. Type 'A' would be Hurricanes and Spitfires inferior to Type 'B' (F.5/34 designs). Type 'C' (F.9/35 and F.10/35) would be the newest available, so that a flow of new aircraft would maintain technical advancement. At no time would squadrons simultaneously be using obsolescent or obsolete fighters. The assumption was that ten overseas squadrons would equip, or be partially equipped, with two-seat fighters. If 'A' Types revealed features desirable in their replacements, these could be incorporated in the F.5/34s, bearing in mind their day fighter limitations. In reality 'A' Types proved too good to replace.

The feasibility of eight-gun arrangements was questionable when F.5/34 was drafted. Maximum speed, set at 265 mph, was raised to 275 mph as armament increased to eight guns. The Air Ministry forecast Bristol's F.5/34 reaching 292 mph and the Gloster entry 282 mph, although Mr H.P. Folland, its designer, claimed 302 mph. When Gloster's prototype *K5604*, Mercury IX-powered (all-up weight 5,400 lb), flew in December 1937 it was too late for integration into the fighter programme. Its performance was better than forecast, top speed being 316 mph at 16,000 ft. In 7.8 mins it climbed to 15,000 ft and to 20,000 ft in 11 mins. Bristol, applying lessons from the Type 133, relied upon a Mercury IX, although hoped to replace it with a Bristol Perseus. Their Type 146 of 4,600 lb flew on February 11 1938 and, prior to crashing in May, had reached 287 mph. Neither was superior to the Hurricane or Spitfire, but exhibited better all-round performance than F.10/35 had called for. After all the discussions, specifications and trials, two glittering, mainly private venture, performers had emerged as winners. The task was to build them in profusion.

One other prewar entrant to the interceptor field is certainly worthy of mention, the Martin-Baker MB 2. James (now Sir James) Martin, CBE, a real pioneer, acquired a shed and a collection of World War 1 huts near Denham in which he and a small band of helpers set about designing aircraft. The first was the 1935 two-seat low-wing monoplane, basically of steel tubing for which the company took out a number of patents. They sold an autogyro, then began the

K5604, *first of two Gloster F.5/34 eight-gun fighters* (RAF Museum P5556).

MB 2 upon which £2,665 was spent in 1935. To embark upon this design to F.5/34 using a three dozen work force in a collection of aged sheds was no mean challenge. More amazing, they began to build it in March 1936. A year passed and the project had cost £26,986 whilst the firm's only income had come from selling two wind indicators. Financially undaunted they pressed on with their easy-to-maintain fighter. Angular in shape, it would be easy to produce, its fin being integral with the rear fuselage to which the rudder was attached. Martin wanted a Merlin for his fighter but had to be satisfied for a start with a Napier Dagger III, the engine the Air Ministry tried unsuccessfully to inflict upon the Spitfire. It was not ideal.

Since the MB 2 answered the official requirement, James Martin managed to persuade the RAF to let him test the aeroplane from Harwell where it was rapidly erected then first flown on August 3 1938. Not surprisingly the 'home built' had been rather costly for its maker, £7,477 having been spent in 1937-38 and another £16,147 before it proceeded to Martlesham in November 1938. By then a partnership between James Martin and V.H. Baker had established Martin-Baker Aircraft, now of ejector seat fame. Not surprisingly there was much interest in this *very* private aeroplane. On December 1 1938 Martlesham's initial report commented on many excellent features. Gun installation accessibility was one, even including a platform for armourers to kneel upon. Fuel tanks could be removed quickly from the roomy fuselage, which allowed easy access behind the instrument panel. Unfavourable comment greeted the trousered undercarriage, but improvement would have been easy. Particularly impressive was the attention to small detail, such as inspection doors. The Dagger could be removed faster than a comparable engine from any fighter.

In flight the MB 2 was a different matter, pilots claiming that it was uncomfortable to fly and was a poor gun platform. Lighter, more effective ailerons and a larger rudder were required. If a retractable undercarriage was installed, increased keel area would be needed and a larger tailplane for additional stability during glide-in. Absence of spats and better elevator control might have improved lateral stability. Heavy ailerons made rolling difficult and, when looping, the inefficient rudder was troublesome. Martin-Baker fitted a new rudder and in July 1939 the aircraft was purchased by the Air Ministry as a test

The Martin Baker MB 2 *eight-gun fighter* (Martin Baker).

bed. No possibility of production ever arose, although it was good enough for the company to be given the go-ahead in 1939 for a successor.

In mid-1936 agreements to build many Hurricanes and Spitfires brought the demise of the F.5/34 and F.10/35. Without these, what for the distant future? Any successor needed more engine power and even heavier armament. A cannon fighter was in design and a Hurricane project existed with wing cannon, but the Spitfire's thin wing appeared to prohibit that type's modification. This and uncertain reliability of wing-mounted cannon led to a further generation of machine-gun interceptors.

Hawker began considering a Hurricane successor in late 1936 upon discovering the Spitfire's advantages over their design. Supermarine, possessing a world-beater, was keen to develop its great potential, the task falling to Joseph Smith following the tragic death of R.J. Mitchell at the age of 49 years in 1937. Vickers took over Supermarine in 1938, injecting large amounts of money. With flying boats and bomber interest the company was now very busy. Hawker had only the Hurricane, hopefully the Henley dive-bomber and Hotspur turret fighter. All would have similar, metal stressed wings. Unfortunately, the December 1936 decision that the Henley would carry metal wings first acted adversely upon the Hurricane which had encountered spin recovery problems. So worrying was this, following its discovery by Hawker, that spinning a Hurricane was forbidden. Although dangerous, it could have readily been cured but the firm was reluctant to do so on the pretext that 'modern flying methods make an inadvertent spin unlikely'. Although the Director of Technical Development agreed, the Director of Training pressed for a solution. Consequently, the DTD informed the firm that a cure must be found. An impasse developed before the RAE arbitrated, the deadlock remaining until October 1937 when delivery of the RAF's first Hurricane was imminent. Farnborough's wind tunnel tests led to a trial installation of a larger rudder and a fillet below the rear fuselage, but 101 Hurricanes were built before this was introduced on the production line in February 1938.

Another Hurricane delay came with the decision to fit Merlin IIs in production aircraft. This mark had revised cylinder blocks causing cowling and fairing lines to be altered. On take off the engine gave 880 hp, 1,030 hp at 16,250 ft, and weighed 1,375 lb. Further small modifications to the air intake, propeller, engine mounting, etc, also brought delays.

Chapter 3

The Sabre awakes

During this frustrating period exciting news reached Sydney Camm. Since the end of 1935 Major Halford of Napier had been working upon a 36.65-litre, 2,000 hp sleeve valve H-section in-line engine, the mock-up of which Camm inspected in February 1937. Rapidly it became the most important engine under development, the Air Ministry setting great store by it, staking much upon its success. Horizontally set cylinders had access above and below to ease maintenance and sleeve valves were being used for the first time in a liquid-cooled engine.

No Air Staff call existed for a new interceptor, so Hawker wrote, on March 12 1937, to the DTD seeking advice upon the most suitable project on which the firm could devote attention. A week later the Air Ministry sent a list placing a new single-seat fighter second and on April 19 1937 — the day that the first Merlin II with ejector exhausts was wedded to the first production Hurricane *L1547* — design work on a new fighter commenced at Kingston, to be powered by the mighty Sabre. Camm's ideas quickly crystallised around a 12-gun/400 rpg fighter with a wing span of 40 ft and a 200 gal fuel load. Estimated speed was 464 mph, which assumed the Sabre would provide more power than it actually ever delivered. Camm also varied his plans to allow for the use of a new Rolls-Royce 2,000 hp engine, virtually two Kestrels wedded into an X-section engine, and named the Vulture. This, the first Rolls-Royce power plant to depart from vee form, had 24 cylinders and was intended for use in heavy bombers. Because of its layout it was more difficult to instal than the Sabre. Later, a third variation of the fighter was designed for high altitude operations, for which Napier had most responsibility.

Group Captain Oxland, Squadron Leader Bilney and Sydney Camm discussed various fighter layouts in May 1937. June brought a visit to Kingston by Major Buchanan resulting in the submission to the Air Ministry on July 16 1937 of drawings for a Sabre fighter. This engine was forecast to develop 1,920 hp at 15,000 ft. Napier claimed that it would be available in 18 months time and would make 400 mph fighters feasible. Success of the Hurricane and Spitfire, though, caused the Air Ministry to hold back on this new fighter and on August 27 1937 Air Commodore Verney, DTD, wrote to Hawker stating 'I think it best you defer work. An Air Staff requirement is coming, acting upon your design.' Uncertainty thus provoked led to Hawker tendering a turret fighter to F.11/37, although Captain Liptrot raised the firm's spirit by re-iterating that a new fighter would be needed.

Part of the Air Staff's concern was attributable to the production delays surrounding the Spitfire and Hurricane. Nevertheless, they were fully aware of the need for a more advanced design and decided to include it in the 1937 experimental aircraft programme. Indeed, the Operational Requirements Committee had met in April 1937 and, unknown to Hawker, discussed a new fighter, placing more emphasis on speed than heavier armament, more guns and less ammunition. Fighter Command took the opposite view, eventual agreement being for a 12 × .303 in gun fighter with 500 rpg. It must achieve 400 mph and have a two-speed supercharger ensuring that speed below 15,000 ft would be as high as possible prior to any additional boost. Service ceiling needed to be 35,000 ft in this aircraft for world-wide operations and it should be 'capable of making attacks on ground troops'. This latter feature was as unique for the period as it was an omen for the future. Hawker lost no time, drawings for a Vulture-engined prototype being complete in September and plans for a mock-up on October 5 1937. Discussion ranged over the question of whether the 'N' (Napier) version should come first or the 'R' (Rolls-Royce).

September's draft specification was considered in depth at a meeting, chaired by the DCAS, Air Vice-Marshal Richard Peirse, held on November 29 1937. Before them were ideas for a 'high-speed single-seat fighter to operate world-wide and replace the Spitfire and Hurricane'. It needed to be faster than likely bombers and combine high speed with density of fire from fixed forward-firing machine-guns in the wings. Additional was a rider that 'satisfactory fire density cannot be obtained by less than 12 guns, the minimum to be mounted'. Nose guns in a twin-engined layout were thought to have disadvantages including the torque to be overcome. Speed at 15,000 ft was initially demanded as 420 mph but later lowered to 400 mph. Climb would remain secondary to speed, duration

Gloster Gladiator Is of 73 Squadron. K7965 was used by the squadron from June 1937 to May 1938, and K7985 from July 1937 to August 1938 (RAF Museum P12012).

be about two hours, take off to clear 50 ft no more than 400 yds, landing be accomplished in 600 yds. Already Hawker had claimed a speed of 464 mph at 20,000 ft for the 'N' version and 428 mph at 19,000 ft for the 'R'.

Much of the draft was agreed, including ground support capability using machine-gun fire. 'Spoiler flaps' seemed a possibility for speed reduction when attacking slow bombers.

These important decisions were a backdrop to the commencement of Hurricane deliveries, *L1547* making its first flight on October 12 1937. By the end of November six Hurricanes were complete and, on December 15 1937, the first of the monoplane fighters of the 1930s, *L1548*, touched down at Northolt for 111 Squadron. Six examples had been delivered by the end of December, but, into 1938, biplanes predominated. The first Gladiators entered service on February 16 1937 and six days later 72 Squadron re-formed to operate them from Tangmere. In April 1937 No 54 Squadron, Hornchurch, became the first to have Browning gun Gladiators and, by the end of May 1937, 14 squadrons of Gauntlets formed the backbone of an interim fighter force. When the year ended nine squadrons had Gladiators, ten had Gauntlets, three still had Furies and, of six development Napier-Sabre engines, one had completed a two-hour run.

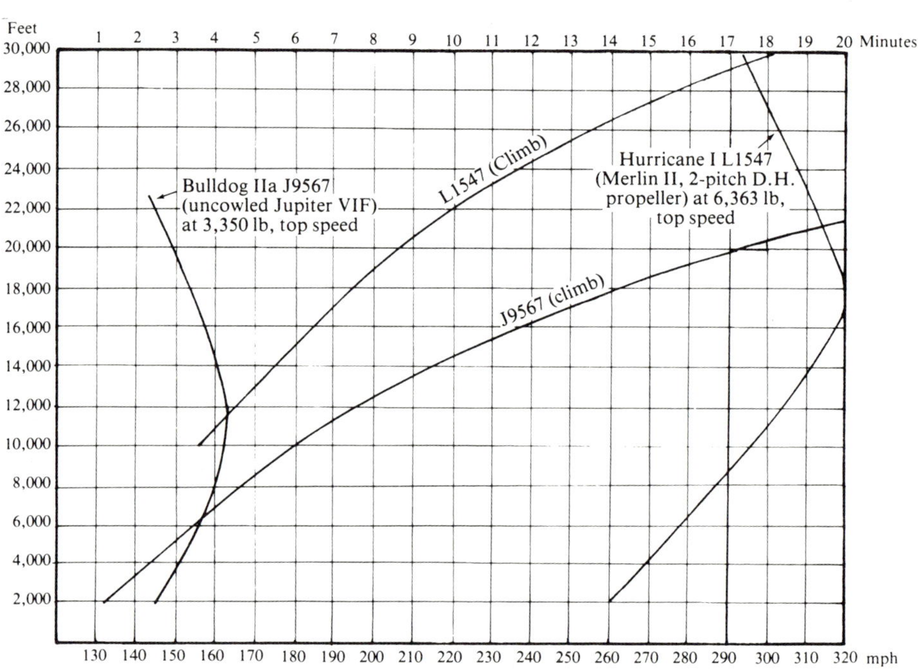

Comparison of top speed and climb of the Bulldog and Hurricane.

Details of the scheme for the new Hawker-initiated fighter, F.18/37, were passed to Bristol, Hawker/Gloster, Supermarine and Westland in January 1938. The Secretary of State for Air was expressing disappointment at the production rate of Hurricanes and Spitfires, the latter delayed awaiting delivery of improved Merlins. Not until late May 1938 did the first production Spitfire

Above *First camouflaged production fighter, L1547 was also the first production Hurricane. Used by Hawker, RAE and A&AEE for trials, it joined 312 Squadron on August 31 1940, and crashed in the River Mersey on October 10 1940.* **Below** *Hurricanes at Northolt in 1939. L1684, nearest, belongs to Station Flight and 'TM-B' to the first Hurricane squadron, No 111 (RAF Museum P2639).*

fly, and a second one on June 17. The aircraft was more difficult to build than the Hurricane, whilst the latter's weight was causing concern since it was now around 6,000 lb. A boost to the Hurricane's public image came in February 1938 when Squadron Leader John Gillam flew one from Edinburgh to Northolt,a strong tail wind giving him an average speed (IAS) of 408 mph — 325 mph TAS at 17,000 ft.

Formal DTD approval of F.18/37 was given on March 9 1938 encouraged by the news that, during a 50-hr type test, a Sabre had given an output of 2,050 hp. Napier reckoned its engine would be off tests in 1938, into production late 1939 and being delivered at a rate of about 30 units a month by December 1940. An

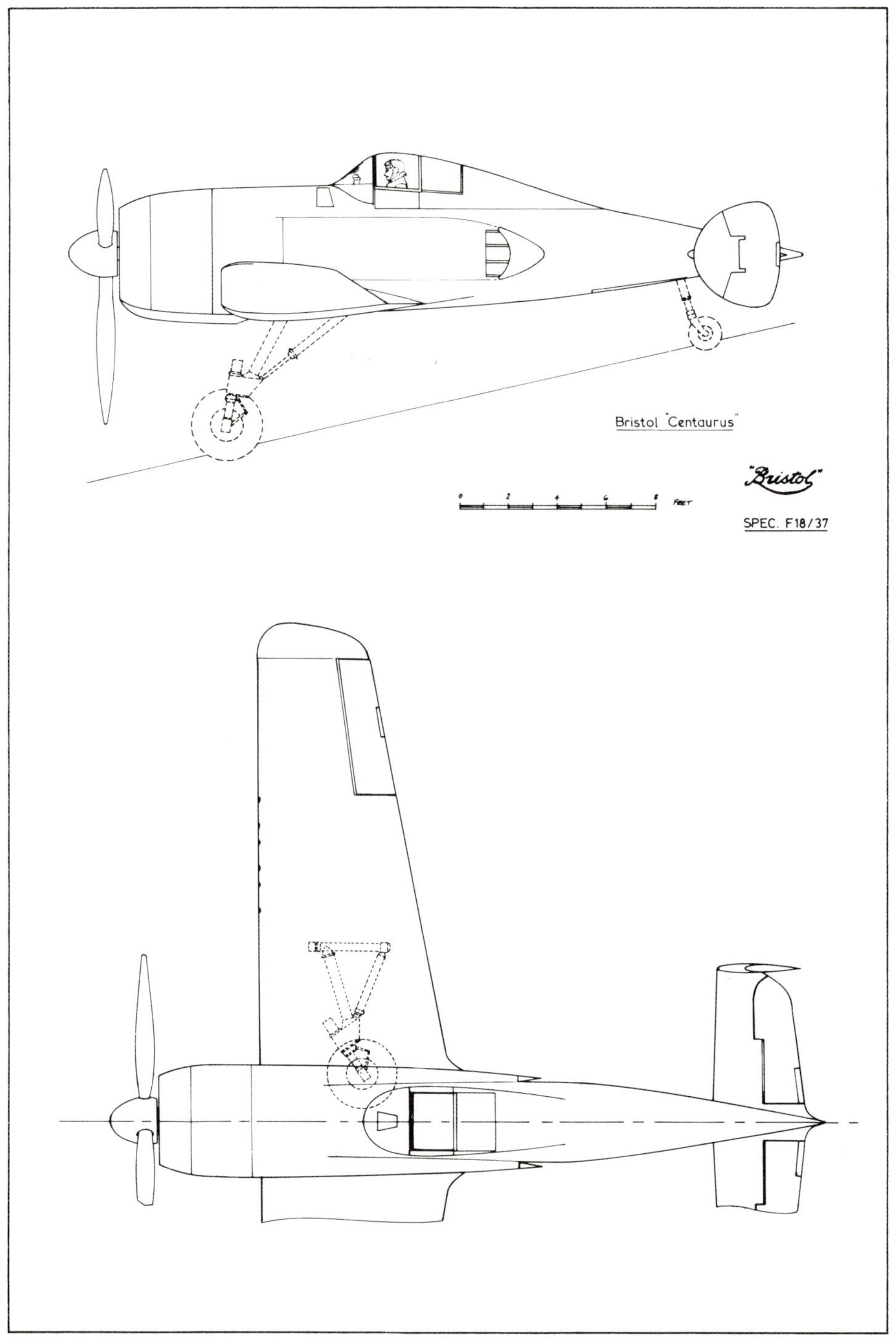

To the Specification which resulted in the Typhoon, Bristol devised this design, powered by a Bristol Centaurus (D.C. Greenman).

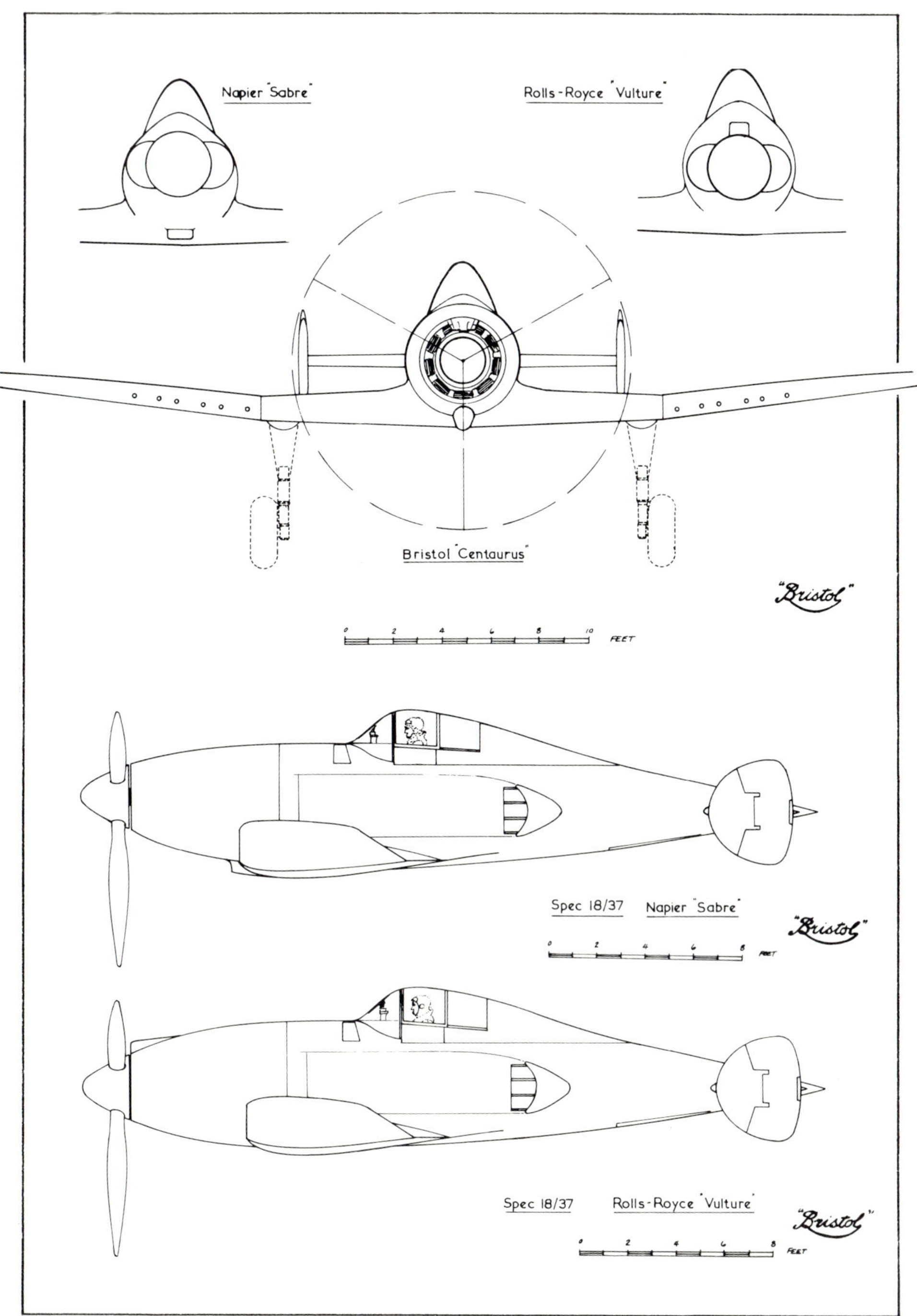

Further versions of the Bristol F.18/37 would have been able to have been powered by either a Napier Sabre or a Rolls-Royce Vulture as alternatives to the Centaurus (D.C. Greenman).

Air Ministry officer reported that the test engine had 'run well'; indeed, it had run deceptively well. Tenders for F.18/37 were invited on March 29 1938, Hawker's submission of April 30 being considered the best. Five companies had devised diverse layouts by May, Westland and Supermarine choosing twin-engined designs, their surprising choice being suitable for tractor- or pusher-propulsion. The Supermarine design, Type 327, was redesigned as a cannon fighter (see page 00). Other layouts were single-engined, Bristol's submission having a range of possible engines while Gloster's pusher layout was reckoned better than Supermarine's.

Early setbacks plagued the Vulture engine, both formidable and very technical. Its availability for bombers by March 1941 was unlikely, so Handley Page set about redesigning their Halifax in July 1937 to have Merlin Xs. This news did not augur well for Hawker's 'R' Type. There was also concern over whether its structure could even accept the very powerful, heavy engines, so both variants were tried in the RAE wind tunnel.

With no desire to have unconventionally laid out designs, the Air Staff narrowed their F.18/37 choice to the Hawker layouts, comparing them with the Supermarine tractor design which could so easily have four nose cannon and possibly a cannon in each wing root. As the accompanying table shows there was not much to choose between these except that Supermarine's scheme needed two of the precious Merlin engines, the others only one engine each.

	Supermarine	Hawker 'R'	Hawker 'N'
Engine(s)	2 Merlin	1 Vulture II	1 Sabre
Span, ft	41	41.5	41.5
Wing area, sq ft	268	244	244
Wing loading, lb/sq ft	42.4	41.8	41.8
Operating weight, lb	11,350	10,200	10,200
Maximum speed at 15,000 ft, mph	432	428 (at 19,000 ft)	430
Take off to clear 50 ft, yd	370	315	315
Landing over 50 ft, yd	724	775	775
Service ceiling, ft	?	37,700	37,200

The ultimate impact upon Supermarine's future of the Air Staff's decision to reject their submission was doubtless not realised at the time. With hindsight it can be seen as a watershed. Not until 1954 would a new type of Supermarine fighter enter RAF service and then as a stop-gap design. Albeit the Spitfire was a brilliantly developed machine with a long, proud Service career, but Hawker's future was about to be assured, even to the present day. Supermarine's reputation would stay with the unbeatable Spitfire.

Before a final decision was taken, the Chief of the Air Staff suggested, on June 12 1938, that details of the performance of Westland's cannon fighter should be studied further for it was due to fly in August. There was discussion of re-engining that aircraft with Rolls-Royce Exe, E.108 or Bristol Taurus engines, and of bringing it in with the less vulnerable and still much favoured air-cooled engines, thus freeing the more powerful power plants for bombers. But, as the F.37/35's first flight date slipped away, the CAS decided instead, on June 27, to confirm the order for two prototypes of each new Hawker fighter. The possibility of it having cannon armament was, raised by the Air Defence Committee in August, following Camm's earlier suggestion, but all cannon pro-

Spitfire I, K9845, joined 41 Squadron in January 1939 and was badly damaged when landing at Catterick on April 3 1939 (RAF Museum P12087).

duction was earmarked for F.37/35. Finance was authorised for F.18/37 prototypes on August 19 1938 and detailed design was discussed at the end of that month. By then the RAF had received its first production Spitfire when *K9792* touched down at Duxford for 19 Squadron on July 30 1938. Re-equipment was painfully slow, only two more Spitfires joining the squadron in September 1938 as the country passed through the Munich crisis. First delivery to the companion 66 Squadron, Duxford, came on October 31, but by the end of the year only 44 Spitfires had reached the RAF.

On September 1 1938 the Air Ministry asked Camm to consider the possibility of a third variant of his fighter, using the forthcoming 2,000 hp Bristol Centaurus air-cooled radial engine. Camm immediately replied that extensive structural modifications would result in a poor view for the pilot. Had Hawker known just what a burden the Sabre would become they would surely have opted readily for the successful Centaurus. Instead, they were pleased to receive on October 20 1939 the third mock-up Sabre for their fighter, followed on December 23 1938 by the 12th Vulture for the first 'R' Type, *P5219*.

By the end of 1938 195 Hurricanes were in service. A major advance had been the fitting of a DH three-bladed, two-pitch propeller on *L1562* in August, by which time the Hurricane's normal all-up weight had risen to 6,300 lb. So ruggedly successful was the Hurricane that on November 1 1938 a further 1,000 — extensively sub-contracted to Glosters — were ordered. December 1938 brought consideration of the fitting of pilots' armour, despite its weight. The armour was added slowly, only 70 Hurricanes having it by June 20 1939 by which time the modification rate was 15 aircraft weekly.

Hawker's own Hurricane, *G-AFKX*, first flew on January 24 1939 featuring a Merlin III. Improved engines were an obvious route to countering rising weight. Another was improved fuel, especially since the Air Ministry was devoting attention to reducing take-off runs. A ten per cent improvement was expected by using 100 octane petrol, experimentally tried since 1934. An estimated 540,000 gal would be needed for Hurricanes in use by September 1938, at a cost in the currency of those days of £23,625 above an equivalent amount of 87 octane spirit. Advantages from 100 octane petrol were well worth that amount, improving power output without increasing weight. This would supplement improvements from the Hurricane's stressed metal skinned wings, the first pair of which became available for a production aircraft in March 1939 and, fitted to *L1877*, first flew on April 28 1939. Another advance was the application of the constant-speed propeller. Combined with higher octane fuel it much improved field performance as these figures show for take off over 50 ft:

Fixed wooden propeller, 87 octane spirit: Hurricane: 563 yd, Spitfire 690 yds.
Fixed wooden propeller, 100 octane spirit: Hurricane 500 yd, Spitfire 610 yds.
Constant speed propeller, wooden, 100 octane spirit: Hurricane 310 yds, Spitfire 355 yds.

Fitted with a two-pitch DH three-bladed 11.03 ft diameter metal propeller, first tested on *L1562* in February 1938, the Hurricane's all-up weight was 6,363 lb. Tests in January 1939 showed such a machine to have a top speed of 320 mph at 17,300 ft and with 80° flap its landing run over a 50 ft screen was 315 yds.

By the end of February 1939 292 Hurricanes and 91 Spitfires were in RAF hands, allowing the release of 24 Gladiator Is to bolster neglected Middle East defences. Prior to shipment they were modified at 27 MU Shawbury, being tropicalised to Specification F.36/37. Time had also come to bid farewell to the Hawker Fury, both No 1 Squadron's Mk Is and No 43 Squadron's Mk IIs having left front-line service in November 1938. Furies were subsequently used as training aircraft.

Commitment of £500,000 for jigs and tooling along with £3m for F.18/37 production was made on December 2 1938, mock-ups being inspected on December 16, followed by confirmation of prototype orders for two of each type on February 14 1939. Construction immediately commenced to meet the first planned delivery in February 1940, the next two to follow within three months. On February 15 1939 the first Vulture flight engine, heavier than forecast and giving less output that predicted, was ready for the fighter. A week later two sets of wings were in the jigs. Although there was little difference between the 'N' and 'R' airframes, the Vulture proved more difficult to install so the fuselage had been enlarged to allow for either power plant. July 1939 brought major decisions. A Vulture had been installed in one airframe, but the Sabre was not yet available. Delays to both versions seemed inevitable so on July 4 another 100 Hurricanes were ordered. Napier had already agreed to produce 100 Sabres and 100 sets of parts, visualising production starting in June 1940 and rising to 300 Sabres a year from Acton and the 40,000 sq ft of factory space leased at Park Royal. Expectation remained that the Sabre would soon advance fast.

To boost production, in November 1938 Napier had drawn up plans for a vast new factory to build 1,000 engines a year. On July 12 Major G.P. Bulman, Director of Engine Production, said that he expected the Sabre to displace the Vulture because Rolls-Royce already had a huge workload. If the Vulture did not improve, then Rolls-Royce Glasgow and part of the Crewe works building the Vulture could be released for Sabre production, removing the need for a new Sabre factory and releasing its work force. The Supply Committee decision of July 4 1939 to order 1,000 'Camm fighters' for delivery by December 1941 brought forward the question of which version to order so that tooling could commence. Suggestions were for 100 Vulture aircraft, then 100 with Sabres, because the former engine had flown 30 hrs already, the latter none. Already, having two power plants showed advantages and drawbacks for it split the design and development teams at Hawker.

On April 12 1939 Supermarine also received an order for 1,000 Spitfires.

K9793, *the seventh production Spitfire I, was used at A&AEE for equipment trials and for DH three-blade CP propeller trials in 1939. After a short spell at RAE it joined 92 Squadron on September 1 1940 and was destroyed in action on September 11 1940* (Supermarine, via Bruce Robertson).

These were to be built at Lord Nuffield's Castle Bromwich factory instead of Whirlwinds, and the designation Spitfire Mk II was allocated to them on June 17 1939. On July 10 the Air Ministry decided to order 500 of each F.18/37 variant, the Vulture version now to be built at A.V. Roe's Yeadon works, the Sabre type at Gloster's. On July 20 1939 the first Vulture prototype, *P5219,* was completed and conveyed to Hawker's new Langley airfield on July 31. By August 1939 the characteristics of the two fighters differed so much that they could not be covered by mark numbers. The Vulture version was named 'Tornado' in August and on September 6 1939 the name 'Typhoon' was given to the other.

One reason for the three-bladed metal propeller on the Hurricane was the need to provide forward balance against the additional weight, arising when the Ministry ordered blind approach equipment to be fitted in the rear of the aircraft. Three-bladed propellers, however, brought many advantages to the Hurricane and Spitfire. The 78th *et seq* Spitfire had three-bladed two-position propellers and from the 175th aircraft a Merlin III was fitted (Merlin II modified for cs propeller), its drive shaft able to accept a DH or Rotol propeller. A dozen trials Hurricanes had the latter in June 1939, the month when Hurricane *L1856* became the first with a Merlin XII engine. Flight trials of a Hurricane with a DH two-pitch propeller began on July 14 1939 and since May 17 *L1699* had been testing tropicalisation equipment.

By the summer of 1939 Fighter Command had sufficient experience to assess the qualities of the in-service Hurricane and Spitfire. Concern had mounted because the in-service performance of both types was below that obtained during acceptance trials*. In a carefully prepared experiment, 12 typical aircraft flew in pairs, each Spitfire 300 yds behind a Hurricane. At a signal from the Spitfire pilot both men opened their engines to maximum revs held for five mins. The Spitfire pilot recorded the time needed to overtake the Hurricane and both pilots noted their speeds, at 10,000, 15,000 and 18,000 ft. Test results convinced Fighter Command that the Hurricane's speed had been much over-

* An early production Spitfire I during A&AEE trials showed a top speed of 362 mph at 18,500 ft; fast cruise being 318 mph at 15,000 ft. Its loaded weight was 5,819 lb. The aircraft climbed to 15,000 ft in 6.5 mins and to 30,000 ft in 22.4 mins, and had a service ceiling of 31,900 ft.

Above *Three Gladiators of 605 Squadron Auxiliary Air Force, shortly before hostilities. N5583 leads N5585 (closest) and N2312 all of which saw lengthy war service* (RAF Museum 5877-24). **Below** *Spitfire K5054 in 1938 style camouflage. It was damaged beyond repair when landing at Farnborough on September 4 1939* (Supermarine, via Bruce Robertson).

rated, whilst that of the Spitfire was even more disappointing, although the latter had insufficient time to work up to full speed when overtaking the Hurricane, which was starting at nearly its maximum speed. So disturbing were the test results that repeat trials were ordered to take place by July 25 1939, for then a report had to be sent to the Under-Secretary of State for Air. Submitted average figures for the fighters' top speeds were: 10,000 ft — Hurricane 288.6 mph, Spitfire 339.3 mph; 15,000 ft — Hurricane 302.1 mph, Spitfire 340 mph; 18,000 ft — Hurricane 302.8 mph, Spitfire 344.1 mph. The reasons for such

All that remains alive of biplane days, the Shuttleworth Trust's Gloster Gladiator L8032 providing, still, the gorgeous shape and sound of the 1920s and 1930s—and even the 1940s, for Gladiators served until 1945 in front line style as meteorological reconnaissance aircraft.

results varied and were, in part, due to ever increasing weight without additional compensating power. Fortunately they did not give a complete picture of the aircrafts' combat capability, and ignored their pilots' skills.

What mattered more at the end of August 1939 was the number of fighters in service and the availability of pilots — along with levels of readiness. On the tragic day when hostilities commenced the RAF's single-seat fighter force comprised the following aircraft: Hurricane — delivered 475, already off charge 75; Spitfire — delivered 305, already off charge 27; Gladiator I — delivered 210, off charge 38; Gladiator II — delivered 224, off charge nil. Four of the Hurricane squadrons were earmarked for Field Force France. Of four overseas squadrons equipped with Gladiators, three held 77 Mk Is and one had 16 Mk IIs. There were 125 Gladiators in Egypt, 37 in Iraq and 16 Mk IIs in Aden. Of the remaining 13 Gauntlet Is, 11 were in the Middle East and of the remaining 125 Mk IIs, 31 were also overseas.

Backing the Regulars were 14 Auxiliary Air Force squadrons, three of which had Hurricanes and another three having Spitfires. Between them three AAF squadrons held 52 Gladiator Is whilst two others had 20 Mk IIs. Seven AAF squadrons held 131 multi-seat Blenheim If long-range fighters, of which 99 were serviceable when war began. No 616 Squadron, non-operational, had 14 Gauntlet IIs against its establishment of 14 + 5 aircraft.

Fighter Command had an establishment on September 3 1939 of 30 single-seat fighter squadrons. Their intended strength was 16 IE + 5 IR, a total of 480 IE and 150 IR aircraft. Total actual strength was 570 aircraft, nine of which were waiting to be struck off charge and 30 of which where under repair. For the 659 pilots posted on to squadrons, 486 serviceable fighters were available. Sixteen squadrons had Hurricanes, ten had Spitfires and three flew Gladiators. No 605 Squadron held a mixture of Gladiators and Hurricanes and No 29 had Blenheims and Hurricanes. Two AAF squadrons were non-operational and equipped with Hinds (610 Squadron) and Gauntlets (616 Squadron).

Chapter 4

We must improve

Whilst it is untrue that Germany lost the war by ineptitude, some enemy policies were foolish. Not to have attacked British airfields at the outbreak of hostilities, even if that meant neglecting the Polish campaign, was a major blunder, likewise failure to attack British factories. Instead, during the first few weeks of war our home-based fighter force improved, expanded and placed itself effectively on a war footing. Aircraft production increased unmolested and plans for future fighters proceeded little disturbed. Obviously, the brunt of any fighting would long be carried by Hurricanes and Spitfires. Already it was certain that the Hurricane would be superceded by the Hawker F.18/37, whereas Supermarine's future lay with the amount of refinement possible with the Spitfire. Much hinged upon Napier finding a base for their Flight Installation Section and, on September 3 1939, the Director of Engine Production visited Luton, the site selected for Sabre flight tests. Leicester had, in August, been suggested as suitable for a Napier factory because of high unemployment in the area. But on September 28 a Lancashire factory seemed a safer siting. Plans were therefore drawn up during October to erect a factory near Liverpool, able to produce 2,000 Sabres a year.

Martin-Baker, whose first fighter had impressed by its ease of maintenance, had for three months held a contract for a new machine, the Rolls-Royce Griffon-engined F.18/39. Its specification stated categorically that Rolls-Royce would supply the engine, its mounting and cooling system — all for fitting forward of the engine firewall. They forwarded details of engine weight, size and stressing prior to the war. Then, on September 15 1939, Martin-Baker received disturbing news. For what had become the first British six-cannon fighter the power plant would weigh 264 lb more than forecast and be longer than advised. A blow had been dealt at what might, just might, have become an excellent fighter. Instead, its entire design had to be changed. The wings were shifted forward and nearly all detail work needed repeating. Yet the contract for its completion by February 15 1940 remained unaltered, likewise the £28,000 allowed for the job. Thus the non-delivery penalties facing the firm were grossly unfair. Second and third prototypes were demanded close on the heels of the first and for £18,000 each. Payment would be reduced for late delivery. Total cancellation without liability was threatened should the first machine be incomplete on time. Yet in September 1939 the Griffon engine was very, very far away — as the Ministry well knew. In desperation Martin-Baker turned from Rolls-Royce to Napier and acquired a fresh load of trouble.

Dependable Hawker, meanwhile, was doing well. A tentative order for 700 Hurricanes (reduced to 500 on November 22 1939 to permit more F.18/37 work) was placed on October 5 1939, the day before the Tornado prototype — all-up weight 9,127 lb — made its first flight. On the strength of the Tornado's early performance, the production order of July 1939 was, on October 14 1939, revised. Originally 500 of the 1,000 F.18/37s Hawker built were to have Vultures, this being the more readily available engine. Another 250 with Sabres would be Gloster-built, leaving 250 to be decided. Alterations to contract now placed half the production batch with Gloster, the rest to be decided later. On October 20 1939 Gloster flew their first Hurricane, *L2020.*

By then the eight-gun fighter had drawn first blood. On October 16 1939 six Spitfires of 602 and 603 Squadrons were scrambled to engage nine Ju 88s of KG 30 attacking naval ships near Rosyth. The Section, led by Squadron Leader E. Stevens, Commanding Officer of 603 Squadron, and using Spitfires *L1050, L1061* and *L1070,* shot down a raider into the sea off Port Seton. Three pilots of 602 Squadron destroyed another off Crail. To Pilot Officer P.W.O Mould of 1 Squadron, operating from Vassincourt in France, went the first Hurricane success. Flying *L1842* on October 30 1939 he destroyed a Do 17 at 18,000 ft near Toul. A month later this historic Hurricane was replaced. After modifications it served with 310 Squadron at Duxford from September 7 1940, to November 2 when it caught fire and crashed on the Isle of Sheppey.

Much revised in appearance by having its radiator bath slung beneath the engine to reduce buffeting, the Tornado first flew in revised state on December 6 1939. That day the possibility of the F.18/37s having wing-mounted cannon was discussed, with the first Sabre to hand, No 95009, for installation at the end of the year in the Typhoon prototype *P5212.*

Meanwhile, harsh conditions in France were proving the Hurricane's stamina and brought a further Gloster order, for 202, on December 5. However, the aircraft's future became insecure when, on December 12 1939, the Air Staff decided that the Spitfire's superiority was such that it would have development priority. What could have been a disaster for Hawker was ultimately a blessing

Initially, the Tornado's radiator was installed below the rear fuselage, but here the bath on P5219 is under the nose. Four banks of exhausts outwardly distinguished it from the Typhoon, and it was quieter in flight (BAe).

in disguise. During 1939 Gloster had delivered 32 Hurricanes and the parent factory 537, making the Hurricane the most numerous RAF fighter, for which armament changes were destined to benefit it most in future. There was little possibility of much increase in its speed and altitude performance, although in an attempt to rescue it Hawker produced a Griffon layout. This meant sweeping forward the wing centre section and the Air Ministry wisely turned down the idea. Had Hawker persevered with air-cooled engines they might well have been on to a winner for ground attack purposes. Instead the company, realising the Hurricane's advantages, on January 12 1940 sent a sketch to the DTD proposing six guns in each wing, which the aircraft could readily accommodate. This, they pointed out, its 'competitor' could never achieve. The desirability of so many guns seems never to have been questioned, it merely followed from the F.18/37 ideas. The Air Ministry agreed to the scheme, ordered a trial installation on February 9 1940 and forecast a handling weight of 6,700 lb which was barely acceptable for the Hurricane Mk I.

Rolls-Royce were equally busy developing the Merlin. Two-speed blowers on the radial AS Tiger VIII and the Pegasus were proving very effective and a Rolls-Royce type was tested on a Merlin X in May 1938. This conferred a considerable advance, but by the time it seemed likely to be ready for service the 100 octane Merlin XX would also be ready, so the Merlin X was diverted to bombers. On April 19 1940 the Air Ministry agreed to the general introduction of 12 guns — but only on the Merlin XX powered Hurricane Mk II. With the decline in importance of the Hurricane Sydney Camm, in January 1940, again raised the idea of a Centaurus F.18/37, for the Vulture was giving less power than expected and bringing cause for concern. Again Camm was near to producing a winner and relieving the F.18/37 programme of impending disasters. Had the effort to be wasted on the Sabre been applied instead to the Centaurus a first class attack fighter, and perhaps interceptor too, would have evolved. The superb qualities of the Merlin engine with its high power/weight ratio and excellent level of reliability tended to cloud the vulnerability of liquid-cooled engines in tactical support situations, distant at this time but to become of paramount importance.

Hawker Typhoon first prototype, P5212, in February 1940. The fin and rudder were soon increased in area. Heavy cockpit combing meant a poor rear view (IWM MH5791).

January 1940 first brought serious attention to increasing the range/duration of fighters, particularly for overseas use. On February 7 the DGRD stated that the Beaufighter, Hurricane and Tornado needed sufficient fuel for a reinforcement range, cruising, of 1,500 miles. External tanks would be carried, but no ammunition. Both Hawker aircraft could acquire this range, judging by a possible two-tank installation in a Spitfire. Hawker, on March 4, submitted a preliminary long-range Hurricane scheme to the Air Ministry, suggesting immediate trials to test its feasibility to reach the Mediterranean Theatre. The Ministry had in mind 200 sets of long-range tanks and tropical equipment, for which the Hurricane's wing structure and layout again made it more suitable than the Spitfire.

The Hurricane's performance was, undeniably, inferior to the Spitfire's and trials in France proved the Bf 109's general superiority to the Hurricane. Fitting a Merlin IIISM or Mk XX to Hurricanes was suggested by Hawker on March 1, yet even then the machine would remain 30-40 mph slower than the Bf 109 which outclimbed and outdived it but did not, in the opinion of the Commanding Officer of 67 Wing, easily out maneouvre his Hurricanes. In the Wing's opinion combat advantage rested with the Hurricane once the Bf 109 committed itself to a tight fight, in which situation constant speed propellers would confer additional advantages. Even better results would come from the Merlin XX Mk II Hurricane, so a trial installation was ordered on March 12 and a first flight scheduled for June or July 1940. But the Hurricane was now a sideline to Hawker's big, new, noisy Typhoon, the prototype of which had flown on February 24 1940, and which bore a hefty nose radiator akin to the Tornado's. A marked difference between the Tornado and the Spitfire was the former's thick wings. Sydney Camm, aware of the aerodynamic advantages of thin wings, suggested a radical 'thin wing F.18/37' to the Air Ministry on March 5. This was a project well in advance of its time and in any case, the thick wings proved ideal for containing bulky cannon.

Spitfire development over the first six months of war included installation of a cannon into each mainplane and improved engines. *K9834*, in a cleaner form as 'N17' special Merlin II-powered, had reached 408 mph showing what was feasible. At the start of the war Rolls-Royce set a team to work upon the Griffon engine, tailoring it for the Spitfire. They claimed it would give an output of 1,735 hp, providing such a 'Spitfire Mk II' with a top speed of 400 mph, but weighing about 800-900 lb more due to necessary airframe strengthening. But the Merlin 'lobby' at Rolls-Royce was very strong, their successful engine looking forward to a brilliant future. As a consequence the Griffon programme slipped for the time being. Supermarine turned towards the 1,240 hp Merlin XX which, whilst it did not lead to a much faster Spitfire, would improve the climb and high altitude performance on account of its two-speed blower. The Merlin XX version became the Spitfire III, a new Mk II having been agreed to be powered by the 1,150 hp Merlin XII tested in *K9788* in 1939 and to be mass-produced at the Castle Bromwich factory.

Both the Air Ministry and the Air Staff were disappointed with the Spitfire III, the prototype of which, *N3297*, first flew in March 1940*. A larger engine and intended 20 mm cannon seemed likely to take the first Spitfire airframe to its limits (around 7,200 lb), although the prototype III was tested at 6,650 lb. It

*Its forecast speed was 390 mph at 21,000 ft and the intention was for it to carry four cannon.

came to feature smaller ailerons and square tipped wings spanning 30 ft 6 in and having an area of 220 sq ft, which usefully improved its rate of roll. Its landing run was long, and the 'clipped' wings were incompatible with improved high altitude performance. A new radiator system was required, and probably a stronger spar. Without the reduced span, 1,000 examples were ordered on October 24 1940. The Mk III's fate, though, was sealed mainly by the decision to fit Merlin XXs to Hurricanes leaving better things for future Spitfires. What the Merlin XX might do for the Hurricane was clear on March 6 1940 when it was suggested that the top speed could be as high as 360 mph, although that too would have meant a new radiator system, revised cooling and considerable structural strengthening.

For the Spitfire a new Merlin with top blower only, the Mk 40 series, would be developed requiring a revised engine mounting. Supermarine pressed for a low level blower too, but there was concern over the aircraft's weight, the aim being to keep that and the wing loading low. Care in propeller choice was coupled to much experimentation.

Come the spring of 1940 and Britain's fighters proved quite unsuitable for the April war in Norway. Although Hurricanes and Gladiators were transported there by sea and put ashore from HMS *Glorious* to operate from frozen lakes and unsuitable landing grounds in support of British troops around Narvik, they lacked range and necessary operational control. Courage, though, was in no short supply. Plans for long-range Hurricanes were hastened by these operations and on April 16 the Air Ministry agreed to a reduced ferry range of 1,150 miles, making additional oil unnecessary.

An exotic scheme devised with typical British ingenuity in moments of crisis, as was the case in the recent Falklands conflict, was the fitting of twin floats to Hurricanes for use in Norway. On April 24 they were ordered as 'urgent' and two days later a pair earmarked for a Blackburn Roc reached Kingston for testing in June. By the end of April the float requirement was extended to Spitfires with an instruction that there was a 'definite and urgent need to equip some on the highest priority'. This was ordered to apply to Lysanders and Gladiators on May 7, but the ideas owed more to imagination than usefulness.

A handful of Spitfires were fitted with floats, among them this Mk IX (Supermarine, via Bruce Robertson).

Spitfire floatplanes were to appear periodically, but a more valuable item was Hurricane *P3462* which, on May 7, commenced overload trials when carrying two 44 gal long range tanks.

A few hours before the Blitzkreig on France started, Hawker received a fearsome shock. On May 9 the monocoque structure of the Typhoon's fuselage — previously viewed with suspicion — fractured in flight. Coupled with increasing Sabre troubles, this brought an abrupt turn in company fortune, delaying the Typhoon for a year. It could not have come at a worse time, for the pace of the war soon became hectic.

Air Marshal Sir W. Sholto Douglas, ACAST, now proposed that only Hurricanes with adequate performance overseas should be considered for long-range equipment. Initially 25 aircraft, enough for a 16 IE + 9 IR squadron, would be prepared — also 20 sets of floats for Hurricanes, allowing their use from ice-free fjords near Narvik. Maintenance in such cold areas 'could be a problem', he acknowledged. Hawker were then asked for a trial of skis on a Hurricane, but this and the float notion were cancelled on May 19 1940. Neither seemed likely to be of much use in France.

Production of the refined 12-gun Hurricane II (Merlin XX) (to Specification Hurricane II/P2 of July 15 1940) was planned to commence in December 1940, and on June 11, three weeks later, the Mk II srs i prototype, *P3269* made its first flight. Hope remained that an eight-gun Mk II could be hastened into production by August 1940, although a longer engine mounting, Rotol propeller, revised cooling system and stronger undercarriage were desirable features. Except for Merlin IIIs, XIIs for Spitfires and VIIIs for Fulmars, all production was to switch to the Merlin XX in autumn 1940, planned output being 30 in September-October, 120 in November-December then 200 monthly until May 1941 when it would reach 60 per week. Gloster was to build most of the Hurricane IIs.

Without the slightest warning, Whitsun 1940 brought shattering news to the entire British aircraft industry. Nearly all development of new types would cease, production of the tried would be expanded and Hurricane output was ordered to be accelerated. Typhoon development was held up and on June 12 1940 Camm expressed to the MAP his shock at this idea. Hawker pressed hard and at a meeting on July 23 1940 the Typhoon/Tornado programme was re-instated, but only after a tough fight.

Meanwhile, Hawker's own Hurricane *G-AFKX* had flown on June 9 fitted with one of the latest Merlin 40 series engines. The company was told firmly that there was no future for such a combination, but they did not give up easily, proposing on July 6 that Gloster produce such an aircraft from March 1941 and that all Hurricanes should mount 12 guns by the end of 1941. Both suggestions were curtly dismissed. Another of Camm's ideas for Hurricane improvement was increasing ammunition to 400 rpg, to which the Air Ministry retorted that 'the guns would surely overheat and cease to function' — after already agreeing to 500 rpg for identical guns in the F.18/37! Instead, it had been agreed on May 30 that a standard Hurricane fuselage be mated immediately to the 12-gun trial wing for assessment at Boscombe Down. The A&AEE reported on June 21 that inaccessibility of the four extra guns made them barely worth fitting and that it was much better to keep eight and find room for 600 rpg. Hawker nevertheless persevered, and in July their second 12-gun machine, *P3811*, was tested.

June 1940 saw the first Canadian-built Merlin II Hurricanes, ordered on

January 4 1939, arrive in Britain. Not until the 25th example, *325*, was the Merlin III standard. By mid-1940 no Hurricanes were entering active service with fixed-pitch propellers, although some were still used as trainers. It was also in June that the first Spitfire II (Merlin XII) flew at Castle Bromwich, from which vast factory 11,939 Spitfires eventually emerged. Early in June an investigation was carried out into advantages of switching from DH two-pitch propellers in general use on Spitfires to the Rotol constant-speed type. An aircraft of 65 Squadron served for comparative tests, revealing that conversion of the DH two-pitch type to constant speed gave similar results to the other one. Such conversion entailed fitting CSU pipes from the engine to the CSC cockpit controls, taking about 20 man hours per aircraft. On June 16 Fighter Command told the Under Secretary of State for Air that experience showed that only Rotol propellers greatly improved the Spitfire's take-off and climb, manoeuvrability, endurance and ceiling. Hornchurch pilots added that their Rotol propeller aircraft was superior to any enemy fighter engaged.

By the end of June 1940 50 Hurricanes were being modified to carry 88 gal of fuel in addition to the normal 97 gal, the extra weight being carried only by metal-winged aircraft. On June 18, 12 Hurricanes with external long-range tanks were despatched to the Middle East, although five were to suffer fuel pump failures. No 602 Squadron pressed for more tankage in their Spitfires but was told this was not yet possible and that cannon layout proposed for Spitfires made extra tanks a difficult proposition.

So to July and then the Battle of Britain, during which Fighter Command's average strength, between July 10 and October 31, was 1,326 Hurricanes and 957 Spitfires. The battle was fought using mainly Hurricane Is — top speed around 310 mph at 17,000 ft, and Spitfire Is — maximum speed 350 mph at 18,500 ft. Both types exhibited a wide variety of modification states.

The first eight-gun Hawker-built Hurricane IIa srs i aircraft were delivered on August 19 1940. Ten reached 111 Squadron on September 2 1940 and went into action from Croydon next day. Although the first Spitfire II, *P7280*, was posted to A&AEE on June 27, and *P7286* joined 152 Squadron on July 17, the first squadron fully equipped was No 611 at Digby, during the third week of August

A Hurricane I modified for desert fighting. It has an air cleaner and early form of drop tanks.

1940. No 266 Squadron received Mk IIs at the start of September, then 74 Squadron. First operational use of the Mk II came on August 31 during 611's daily detachment to Duxford. The single-speed high-altitude supercharged Merlin XII, using 100 octane petrol and rated as best at 1,150 hp at 14,500 ft, and which drove a Rotol constant-speed wooden propeller, conferred on the Spitfire about 6,000 ft additional ceiling and improved climb. By the end of 1940 354 Mk IIs had been delivered.

There was talk of a cannon Spitfire II, only possible by fitting the new wing planned for the Spitfire III. July 1940's plans for increased armament on the Hurricane II/P2 were also underway. On August 14 it was decided the Mk II/P3 would need a pressurised engine cooling system, be convertible for tropical service, have stressed metal skin wings, 12 × .303 in guns and a Rotol CS propeller. Austin Motors would build this version. By the end of the Battle of Britain, though, the Hurricane was an obsolescent interceptor, its early production lead the main reason why so many were in use.

It was fortuitous that so many fighter aircraft were available. Sensing the strong possibility of a shortage, Phillips and Powis of Reading had, in May 1940, reviewed the situation. Firstly the company devised the conversion of 26 Master I trainers into reserve fighters by the fitting of six .303 in wing guns, a reflector sight, bulletproof windscreen and armour for both pilot and engine. This reduced the top speed from 228 mph at 14,500 ft to a little over 200 mph. The aircraft never went into action. Secondly, in May 1940, the company undertook preliminary work on a simplified wooden fighter discussed with the MAP on July 13 1940. Soon it won Air Ministry backing after Phillips and Powis promised to have a prototype completed in three months' time. Non-essential services were reduced to a minimum whilst fuel and armament would be as plentiful as possible. Hydraulics would be deleted, so the aircraft had a fixed, spatted undercarriage. A Merlin XX 'power egg' as prescribed for the Beaufighter II was chosen, and the eight-gun mountings were standard Hurricane type. Master components were used to speed construction.

The promise was kept, and the Miles M20 marked *U9*, first flew on September 15 1940. Thick wings and hefty undercarriage induced more drag than forecast, but the use of smooth ply skinning and fine paint helped improve matters. Fuselage lines had been designed to avoid double curvature, making skinning easier. All spar joints were deleted to reduce weight, there being two spars and closely spaced ribs. Access doors over the eight guns were large. The clear cockpit bubble hood was very unusual and afforded an excellent rear view. That was a good feature, but even more advanced was the fitting for jettisonable external fuel tanks making possible cruising flights of three and a half hours' duration. So impressed with the design were the authorities that on

Opposite top *A typical Battle of Britain period Hawker Hurricane MkI P3878:YB-V at Debden. It joined 17 Squadron on July 1 1940 and was written off in a crash landing at Debden on September 24 1940 (RAF Museum P10064).* **Centre** *Spitfire I P9386, mostly flown in September 1940 by 19 Squadron's Commander, Squadron Leader Lane. Briefly with 257 Squadron May–June 1940, it operated with 19 Squadron from September 3 to September 26 1940, and was later used by 52, 57 and 58 OTUs.* **Bottom** *Hurricane IIa Z2521 initially served with 249 Squadron in February 1941, soon joined 242 Squadron, moved to North Weald in July 1941 and then to 247 Squadron. It joined 135 Squadron at Honiley in September 1941 and remained there when the squadron went overseas in November 1941. It was passed to the USSR in September 1942 (BAe).*

The Miles M 20 eight-gun fighter.

August 29 1940 Specification F.19/40 was drawn up, covering this 'experimental wooden fighter' requiring a top speed of 'not less than 360 mph at 21,000 ft', a ceiling of at least 32,000 ft and the ability to land, over 50 ft, in 650 yds.

By the time the prototype was flying no need remained for such an aeroplane which, in prototype form, had a top speed of 345 mph. On July 3 1941 Miles Aircraft (Phillips and Powis renamed) was approached to produce a cleaner version for naval shipboard use, for which a test wing carrying 2×20 mm cannon was ordered. RAF interest by then had radically changed and emphasis was being placed upon high-flying fighters.

Any improvement in high altitude performance of fighters was welcome. At the commencement of the Battle of Britain enemy aircraft came in very low facing British fighters whose performance was best at around 16,000 ft. Low level boosting of the Merlin was increased by 40 per cent to $+12$ lb instead of $+6\frac{1}{4}$ lb. This made the Hurricane and Spitfire 25 mph faster for short periods, forcing the Luftwaffe to operate higher, a trend highlighted in October 1940, commonly called 'Messerschmitt month'. Dozens of enemy fighters flew very high over south-east England, taxing the British fighter force to the full. Only fighters fitted with the Merlin XX, rated at 1,400 bhp at 8,500 ft and 1,405 bhp at 14,750 ft and using 100 octane fuel, seemed likely to be able to fight effectively in 1941. Production of that engine was only sufficient for its use in the Hurricane which, Hawker stated, would from about the 350th Mk II carry 12 guns. In October 1940 production of the Hurricane II srs ii started, its slightly longer nose compensating for the heavier armament. Following the fly-off from HMS *Argus* of a dozen Hurricane Is to Malta on August 2 1940, sufficient tropicalised Mk Is were in the Western Desert by October for No 73 and 274 Squadrons to equip with them. On November 2 a requisition for 1,250 Hurricane II/P2s was agreed, each machine to cost £4,000. But it was to the Typhoon that interest was increasingly directed by Hawker — and the picture was far from encouraging.

In June 1940 a Napier Sabre 1 giving 2,200 bhp had somehow slipped through the 100-hour type test — in the month when it should have entered production. Concern was tempered by the knowledge that it had novel features. Inherent risks of over-heating were appreciated, lubrication was imperfect, cylinders and sleeves were liable to deformation. Ministry and industrial experts were not unduly worried and reckoned these problems could be solved. There was already disquiet over Napier's lack of drive, the firm being officially recorded as 'tinkering with the engine's development'. It would run the engine for a few hours then, hopefully, submit it for type-testing. After failure the company would soon re-submit it and with luck achieve a pass not equating regular reliability. Not for some time did the Air Ministry realise what was going on — and only long after they had taken a chance, ordering the Sabre for a substantial number of aircraft. That risk was taken in December 1939 when, desperately needing a 2,000 hp engine, the Air Ministry decided the Sabre must be mass-produced at the new plant near Liverpool. A few weeks later Tornado/Typhoon production commenced painfully and slowly.

Since Expansion Scheme CP 218 of October 1938, including large scale production of F.18/37, neither intended engine had received preference and the Vulture was also in trouble by 1940. For Hawker's drawing office, working on two aircraft simultaneously brought added complication, likewise the mid-1940 order to concentrate upon Hurricane production and pass the Typhoon and Tornado to 'daughter' firms. Delays surrounding F.18/37 airframes tended to obscure the very serious engine problems. Such was the belief in the Sabre's ultimate success that in autumn 1940 strenuous effort was made to sell the engine, or production rights, to America. There was much interest, until the Americans decided the Sabre too novel, differing too much from their engines.

Typhoon and Tornado delivery was, on September 27 1940, estimated as commencing in February 1941, with 80 Tornados and 86 Typhoons in RAF hands by June 1941, after which about 40 of each type should become available monthly.

First indications of Typhoon capability were evident on October 29 1940 from a report following official trials of *P5212* undertaken at Langley, with the aircraft flying at 10,620 lb, which Hawker claimed was really 10,485 lb. Tare it weighed 8,148 lb and its fuel load was 155 gal. Ratings given to the Sabre NS 1 were 1,950 hp at 1,740 ft and 2,250 hp at 14,500 ft giving an FS TAS top speed of 410 mph at 20,000 ft reached in 4.7 mins. The aircraft took 14.3 mins to reach 30,000 ft and stalled, flaps down, at 70 mph IAS.* Cockpit entry was via a car door-type of doorway, heavy cockpit combing giving a blind rear view. Pleasant to fly, the aircraft had no particular vices and seemed to 'get along without any effort'. Psychologically beneficial was its wide speed range. Take-off was reasonably good, but its 32,300 ft ceiling (compared with the Spitfire II's 37,600 ft) was totally inadequate. 'Careful, painstaking development should make it a very effective fighting machine' was the offical verdict and more trials followed in November-December. On December 5 the second Tornado, *P5224*, first flew.

Although the Typhoon production specification was approved on September 23 1940, the RAF would have a long wait for the aircraft. It was indeed

---

*Forecasts in August 1940 had suggested a top speed of 400 mph at 15,500 ft for the Typhoon and 380 mph at 17,500 ft for the Vulture II Tornado.

fortuitous that Rolls-Royce had a new Merlin to hand for the Spitfire, since otherwise the fighter force would have been totally outclassed in 1941. 'If there was any miscalculation regarding the Sabre,' commented an official review, 'it was in the intrinsic value of Napiers, not the Sabre. It is fair supposition that had the Sabre been in Rolls-Royce charge the engine would have matured within reasonable time.' Clearly, engines were becoming the dominant feature in the fighter programme, but the question of armament was also becoming important.

Chapter 5

Innovation — or inspiration?

By 1941 separation of machine-gun and cannon fighters as specialised classes had gone. Origins of heavy cannon fighters are described in Chapter 11, whilst the development of single-engined machine-gun and cannon fighters continues here.

Applying the Merlin XX to the Hurricane and leaving the Spitfire unimproved was unthinkable, so Rolls-Royce investigated the possibility of a low-altitude blower on the Merlin X. That innovation became the Merlin RM5S, soon the Merlin 45. Impressed by its trials, the Director of Engine Production requested performance estimates of a Merlin 45 Spitfire. They revealed an 18 mph speed increase above 17,000 ft, improved rate of climb of about 350 fpm above 16,000 ft and ceiling heightened by 1,000-2,000 ft, compared with the Spitfire II. On Christmas Eve 1940 this forecast was discussed during a major meeting held at Boscombe Down to review the entire Spitfire programme. Unanimous desire was expressed for full scale production of a Merlin 45 Spitfire by April 1941 — provided it did not prejudice the Hurricane II — in the belief that forecast figures would be bettered in reality. The Commander-in-Chief, Fighter Command, requested spare Merlin 45s — which he called a 'priceless asset in the struggle for air superiority in the south-east corner of England' — be installed in earlier Spitfires when on repair lines. But he also expressed, on December 27 1940, general worry over the future, stating, 'We're outclimbed and out manoeuvred by the Bf 109 at 25,000 ft'. That situation he reckoned would remain, even after the Merlin 45 Spitfires came along.

Next day the Ministry of Aircraft Production wanted the Spitfire III immediately cancelled in favour of fast production of 500-600 Merlin 45 aircraft for which, by the end of 1940, an initial contract for 100 was agreed. Spitfire I *K9788* was, in January 1941, hastily fitted with an RM5 among confusion over the interpretation of instructions, as a result of which Rolls-Royce were installing the engines in ex-19 Squadron cannon-armed Spitfire Is by January 15. Within a week the error received a blessing and conversion was sanctioned and hastened. A survey of January 21 revealed 25 such Spitfires remaining out of the initial 30, of which two were with 19 Squadron, three with No 92 and six at AFDU Northolt. All were suitable for conversion. Preliminary rating for the RM5 was 1,205 bhp at 15,500 ft using + 9 lb boost, 1,200 bhp at 3,000 revs for take-off using 12½ lb boost, and maximum output possible 1,250 hp.

More Merlin 45 Spitfire production was planned on January 15 1941. Supermarine would produce 500 between February and September 1941 and the

P8332, *one of the few Spitfire IIbs, became* ZD-L *of 222 Squadron with which it served from May 21 1941 to August 27 1941* (RAF Museum P8706).

gap between Mk II and III production at Castle Bromwich would be filled by Spitfire II airframes mounting Merlin 45s, likely production between April and November being between 200 and 500 aircraft. Merlin 45s would be installed in new aircraft only, older IIs being unsuitable without rewiring. Mk II improvement started in February 1941 when the first of 100 sets of two-cannon (drum-fed) and four-machine-gun wings became available for the Mk IIb.*

Another major Spitfire improvement had come about in November 1940 when it was decided that all Spitfires would have metal ailerons to enhance control. † 'The effect of metal ailerons on the Spitfire are so great one has to fly the aircraft to believe it', commented a test pilot before a trials aircraft, *P9505*, was on December 27 1940 posted to 266 Squadron for testing by five squadron pilots. They agreed on the excellent improvements in the rolling plane, and at all speeds, and that aileron control was now as light as the elevators'.

Early in February 1941 the Merlin 45 Spitfire was designated Mk V, but no suffix letter indicating armament type had yet been officially approved, although the Mk Ib (Spitfire I/P2) had been authorised as long ago as March 15 1940. Supermarine was ordered to build five Mk Vs by the end of the month and by then Rolls-Royce had completed three of their authorised 23. Each needed a new radiator, header tank, air intake, cowling and engine cooling control changes, as well as a revised fuel system. A schedule of Mk V production, devised on February 5, indicated: February — 15; March — 80; April — 130; May — 184; June — 295; July — 265; August — 240 and September — 195. Planned monthly Mk III production from June to October was to be 10, 45, 80, 125 and 30 aircraft respectively. Building sufficient Merlin 45s — 15 in January and 60, 80, 120 in succeeding months — was matched by a Vickers forecast of January 22 for the completion of five Mk Vs in February, then 40, 80 and 120 in succeeding months, all with two cannon and four machine-guns. Vickers had tried such an aircraft, but the opinion of N.E. Rowe, DTD, was that they had

*Production total of the Mk IIb is variously given. An official survey of Spitfire production in 1941 lists it as follows. Supermarine: March — 24; April — 42; May — 21; June — 3; Castle Bromwich: March — 30; April — 40; May — 54; June — 37; July — 7. Total: 258
† How much effect heavy ailerons had upon the Spitfire in combat situations was unappreciated until 1940, by which time it was inopportune to change them. Modified section and various balances were tried until the improvement desired was brought about by metal in place of fabric skinning and a very thin trailing edge first tested on *R6718*.

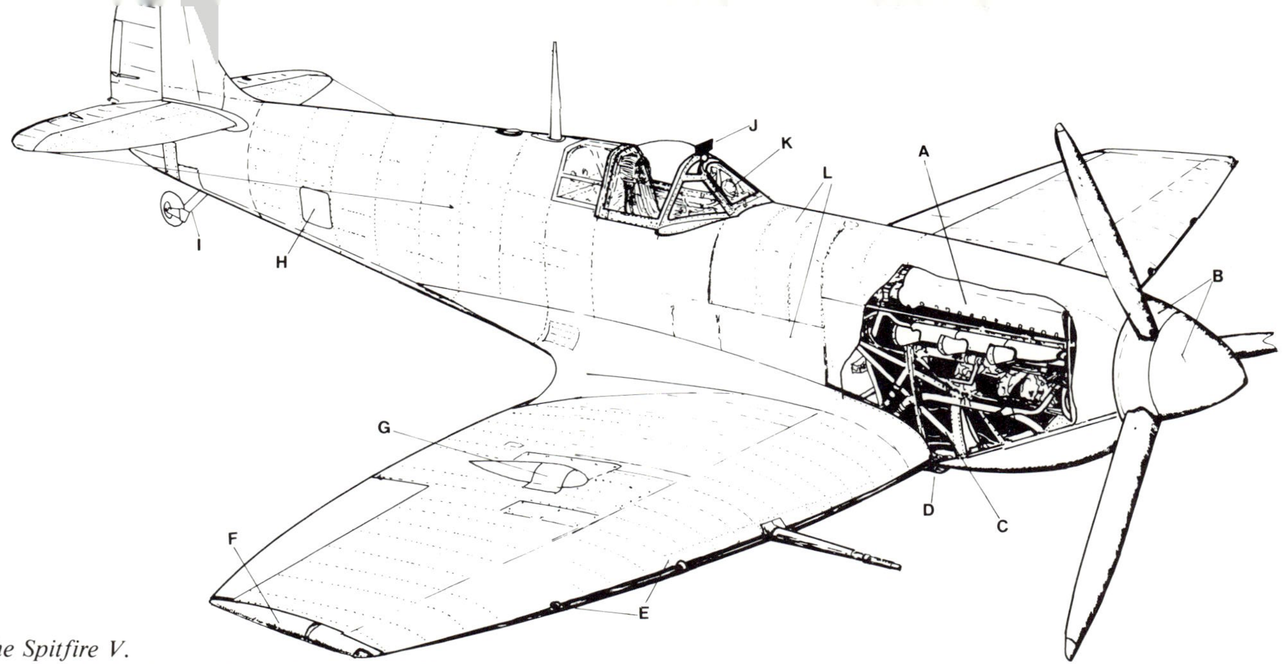

The Spitfire V.
Key: *A — Rolls-Royce Merlin 45, 46, 50 or 50A (Mk F Va, Vb, Vc) or Merlin 45M, 50M or 55M in LF versions, B — propeller, de Havilland 5/39 CS or Rotol RX 5/10, C — engine mounting, D — carburettor air intake (much enlarged to include cleaner on tropicalised aircraft), E — two .303 in machine-guns, in each mainplane, F — detachable wing tip, wooden fairing shown in place reducing* **wing span to 32 ft 2 in** *(normal span 36 ft 10 in), G—Fairing over loading bay of cannon, and access doors for ammunition, H — access door to equipment in rear fuselage, I — non-retractable tailwheel, J — rear-view mirror, K — reflector gun sight, L — fuel tanks, 48 gal upper and 37 gal lower (additional slipper tanks of 30, 90 or 170 gal could be carried below the main tanks. A 29 gal rear fuselage tank was fitted, for ferrying, in conjunction with the 170 gal tank).*

Normal loaded take-off weight of the 8-gun Mk Va was 6414.5 lb, of the 2-cannon 4-machine-gun Mk Vb 6,622 lb. Each machine-gun had 350 rounds, each cannon 60. The loaded Vc weighed 7016.5 lb when fitted with four cannon, or 6,785 lb carrying two cannon and four machine-guns. Aerofoil section NACA 2200 series, dihedral 6°, incidence 2°, wing area (full span) 242 sq ft, fuselage length overall 29 ft 11 in, height one-blade vertical 11 ft 5½ in, tailplane span 10 ft 6 in, tailplane width 4 ft, wheel track 5 ft 8½ in.

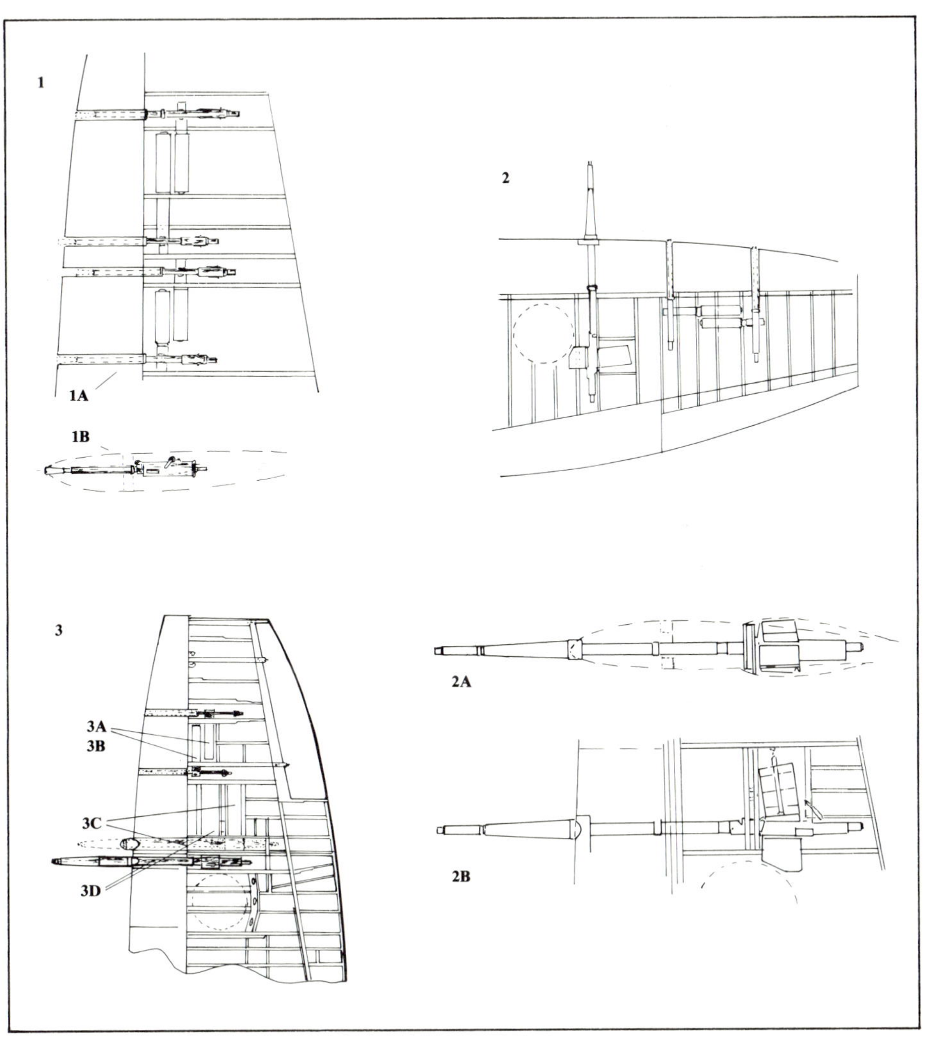

Spitfire armament.
1. *Standard layout for the 8-gun Spitfire Mk I, IIa and Va with the guns staggered to account for their ammunition feed systems. 'B' shows the side view of the gun installation. 2. Layout for the Spitfire 'B' variants. Staggered twin .303 in Browning guns are sited outboard of the single 20 mm Hispano Mk II cannon shown here in a Mk IIb wing. The impossibility of installing the gun without some additional fairing is shown in view 'A'. 'B' shows the plan view of the cannon installation. Bulges above and below the wing have been added over the cannon-loading area 3. By staggering the position of the guns, the 'C' Spitfires starting with the Mk Vc (with revised structure) were able to carry two cannon in each wing. 'A' and 'B' were the ammunition boxes/feeds for two .303 in guns themselves staggered. 'C' shows the position for an outer cannon and rear-placed feed. 'D' shows the inner cannon and feed. Frequently only one cannon was fitted in each wing and the number of guns could be varied. Another version was Type 'E' whrein the inner cannon bay was occupied by a .5 in machine-gun. Outer machine-guns were not fitted. Stubs were frequently seen placed in vacant cannon bay openings and machine-gun ports were variously covered with fabric.*

Above *Canadian Car & Foundry Co-built Hurricanes saw plentiful RAF service. Among them was Hurricane I T9519 (Merlin III) which reached Britain in September 1940, then served with No 312, 315 and 303 Squadrons before joining 239 Squadron of Army Co-operation Command in January 1942. Later it was used by 56 OTU, 2 TEU, 81 OTU, and served with 1665 CU in August–September 1944. It was struck off charge at Peter-borough on October 1 1944 (via J. Robertson).* **Above** *Hurricane IIc Z2905 served as a prototype for extended range trials (BAe).*

no realisation of the urgency of the Mk V and a 'too small experimental depart-ment' to cope.

The Mk V's quality was clear when a two-cannon/four machine-gun example revealed an all-up weight of only 6,460 lb, 200 lb less than the Mk III, which it outclimbed to 18,000 ft. Merlin 45 oil temperatures certainly ran a trifle high and the propeller's CSU needed pitch setters. Otherwise all was encouraging enough for Vickers to go ahead on February 12 1941 with planning production of 520 cannon/machine-gun examples now designated Vbs, the first to be delivered in April. Surplus engines would fit into eight-gun Spitfire Vas built between February–March 1941 before production concentrated on the Vb. The former managed a service ceiling of 39,000 ft — 1,000 ft above the Vb. A temporary expedient had become a mighty successful one.

From cancellation of the Mk III on March 31 1941 was salvaged the 'Universal Wing'. Fixed points for alternative armament, eight .303s or two 20 mm belt-fed Hispano and four .303 in machine-guns or four belt-fed 20 mm cannon were featured, along with possible clipped wing tips, useful later.

Early Supermarine production of the Vs was rapid: Mk Va: 9, 16, 6 and 28 monthly, February–May; Mk Vb: 15, 27, 67, 97 in consecutive months, March–June. From Supermarine there eventually emerged 94 Mk Vas and 780 Mk Vbs, and no less than 3,003 Vbs from the Castle Bromwich works.

What of the seemingly eclipsed Hurricane? A two-seat trainer of December 1940 lapsed, four-bladed Rotol propeller tests were under way and on January 11 1941 consideration of a maritime patrol Hurricane carried pick-a-back on a Liberator was discussed. In January 1941 55 Hurricanes were being tropicalised at Hucknall whilst at ASUs two aircraft a day towards a committment of 170 Mk IIs for the Middle East were prepared after the prototype, *V7480*, first flew on February 8 1941. Hurricane II trials revealed stiffening of controls above 37,000 ft curtailing operational flying. Delivery of the four-cannon Mk IIc commenced with nine in February then 6, 6 and 28 in following months and slowed by the need for cannon elsewhere. Ever-increasing importance was attached to extending the Hurricane's long-range ferry capability and particularly its use as a fighter-bomber.

Chapter 6

High — or low?

In the belief that the future lay with the Griffon, Rolls-Royce were in December 1940 ordered to hasten its development although its power output was little better than the Merlin XX's. By February 1941 Sir Patrick Hennessey of the MAP was pressing hard, suggesting a Griffon Spitfire would attain 430 mph at 23,500 ft and have a ceiling of 38,600 ft. A two-stage Griffon could give a six-cannon Spitfire a top speed of 470 mph at 35,000 ft and a ceiling of 46,300 ft, claimed the makers. Air Staff appraisal credited a Spitfire IV (Griffon) with 410 mph at 22,500 ft and a service ceiling of 36,500 ft — little better than the Typhoon which they had reckoned would attain 424 mph at 22,000 ft. But the Air Staff, believing the Griffon had more future than the Merlin, decided to back its development for intended use in all Spitfires from early 1942. Summer 1941 brought it a boost as the Vulture programme slowed. With the Sabre in trouble, the Griffon looked a feasible successor. Could it replace the Sabre in the Typhoon?

Wrong assumptions concerning the latter were not much in evidence when production began in February 1941. Certainly there was trouble with engine mounting vibration, and a forced landing by the prototype due to engine malfunction. That delayed the development programme and the second machine did not receive a power plant until March 22. Typhoon delivery set at five in April 1941 and intended to rise to 80 monthly in the second half of 1941 was abandoned. It was hoped that the Sabre II running at 4,000 revs might increase the Typhoon's ceiling. With the Sabre and Vulture performing poorly the aircraft's future fluctuated with the engines' favour. Late in 1940 the Tornado production programme was switched to A.V. Roe who were experienced with Vulture vicissitudes. Gloster would build only Typhoons but Hawker probably some of each. Another update of March 1941 required A.V. Roe to produce 351 Tornados and Hawker 58, with Gloster making 891 Typhoons and Hawker 76. Since both engine types were in trouble the RAF was thankful for the Spitfire V. New Hurricanes, too, were backing them.

The first production 12-gun Hurricane IIbs appeared in March 1941 — when Hurricane production reached 12 a day — and were not wholly successful, having insufficient ammunition stowage space. No 605 Squadron flew the aircraft intensively, reporting on March 24 general satisfaction, especially with the 6,000 ft ceiling increase over the Mk I. A specialised Hurricane II intruder with six .303 in guns (150 rpg) and long-range tanks had also been cleared for operational use. On order were 100 pairs of 20 mm cannon wings with another

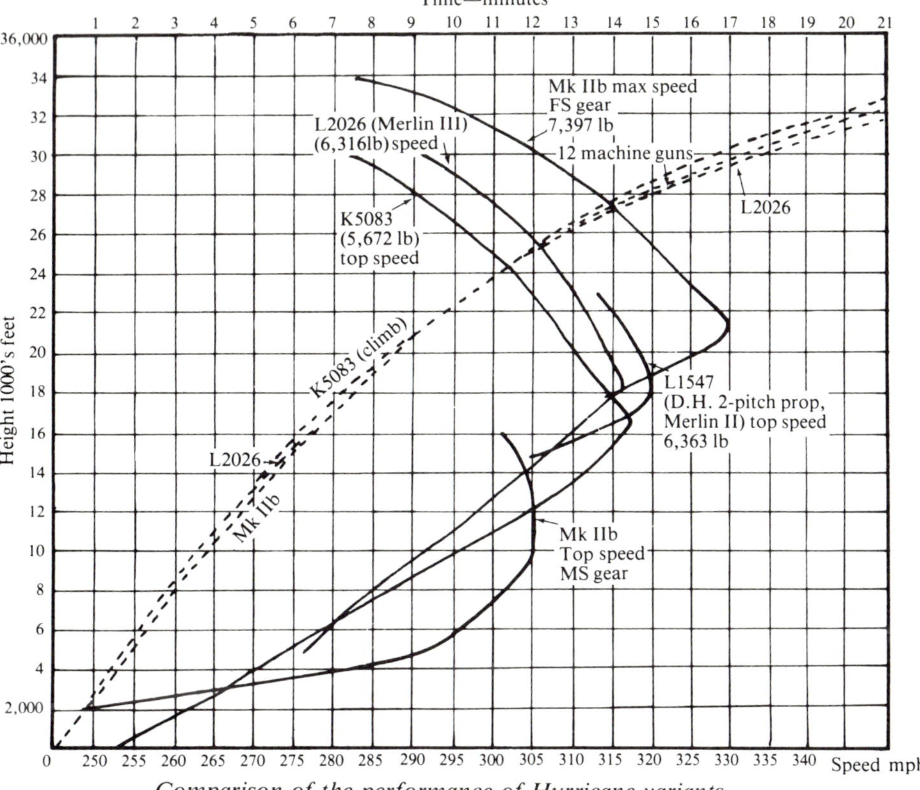

Comparison of the performance of Hurricane variants.

200 about to follow. Cannon production received a boost on April 12 1941 when BSA completed their first Hispano. By September they had built 200.

The Hurricane's suitability for mixed loads in ground attack roles far exceeded the Spitfire's. This was emphasised in April 1941 when the Chief of the Air Staff sanctioned a twin 40 mm cannon anti-tank Hurricane, and turned down Camm's latest Griffon scheme. Authority was given on April 28 for production of 100 40 mm Vickers cannon, for delivery in six months and emphasised the Hurricane's ground support role.

Improved Spitfire armament utilising the Mk III's Universal Wing, and its stronger undercarriage as well as the Mk II's radiator system, was ordered to be introduced in August 1941. Before further advances could be undertaken the Spitfire's airframe needed strengthening. Utilising that, the Type 'B', the four-cannon Mk Vc was born. Universal Wings for earlier marks were cancelled on grounds of weight. Supermarine held large orders for earlier airframes, so stressed their usefulness should fighting revert to lower levels. That forecast was correct and the company must have viewed with concern large contracts for ground-attack Hurricanes. Spitfire improvement was certainly remarkable and often arose from attention to detail. It is well illustrated by *N3053* converted from Mk I to Vb, which, fitted with a four-bladed propeller, attained 40,670 ft and had a top speed of 369 mph at 25,000 ft reached in 8.5 mins. It climbed to 35,000 ft in 14.3 mins.

Modification allowing Spitfires to serve overseas was difficult, though. Additional equipment reduced the speed by about 12 mph, half attributable to the huge air cleaner. Increased fuel was essential and, in any quantity, could only be externally carried, possibly using 44 gal tanks already developed for the Hurricane. Estimates suggested the Spitfire's structure would permit sufficient loading for an amazing 2,800-mile transit. The most a pilot was reckoned able to cope with, in a fighter, was 1,500 miles. Insufficient for ferrying purposes, it led to radical schemes being suggested, for overseas ferrying, in May 1941.

Not until May 3 1941 did *P5216*, the four-cannon Typhoon second prototype, fly, much retarded whilst the Air Staff argued over adopting cannon as standard fighter armament. To hasten Sabre development, an early Typhoon was to join Napier at Luton and on May 26 the value of this idea was reinforced when coolant problems brought down *P5212*, on the day when Gloster flew their first production Typhoon. Concern remained over its structural strength, dives of the Typhoon at 550 mph IAS being ordered as a check. Another inevitable demand came on May 4, for a 'Universal Wing' accommodating six cannon, or two cannon and eight machine-guns, or 12 machine-guns. Hawker had little choice but to accede, signing a contract on July 2. This could only further delay the Typhoon, certainly the Mk Ib cannon version.

May 1941 brought a major appraisal of the fighter programme. In a memorandum of the 10th, the Chief of the Air Staff outlined for the MAP future RAF aircraft needs. 'With fighters, height is gauged important', he wrote, adding that the most promising line of development remained with the Spitfire capable of exceeding 40,000 ft. 'These ceilings', he considered, 'are not an inch too high.' Poor Typhoon and Tornado ceilings and the marked inferiority of the Hurricane when compared with the Bf 109F, might be remedied by fitting blowers to engines. 'At over 35,000 feet, approximately, the human body needs artificial aids', he pointed out. 'Apart from the Whittle jet we have no fighter under design, no unpressurised fighter of higher performance than our present types, due, I think, to us not having an engine of over 2,000 hp. If the Whittle fails our next step is a difficult one. I think we

AA878, a Spitfire Vc four-cannon fighter.

should place under design our next replacement which, I think, must possess a speed in the order of 475 mph at 25,000 ft, and have a service ceiling of 40,000-42,000 ft. The policy of change to 20 mm guns is being richly justified.'

Lord Beaverbrook replied that 'we are alive to the urgency of high altitude fighters, and have arranged large scale production of the Spitfire V, already giving a ceiling of 40,000 ft. The ceiling of the Typhoon will be better than you suppose by the time it appears in quantity. We expect its ceiling to be over 35,000 ft. For a higher ceiling we rely on Sabre development being energetically pressed, and are gradually improving Sabre and Vulture performance at great heights. Apart from the Whittle, MAP is unable to introduce engines of great power during the war. There is,' he added, 'a high priority to improve Typhoon performance aerodynamically.' Amidst unjustified optimism his final statement was one which ultimately transformed the British fighter scene.

To obtain better high-altitude performance, Rolls-Royce was increasing work on the two-stage Merlin 60 commenced in 1940 and expected late 1941.* Limited resources meant that Rolls-Royce could not also cope with the Vulture and Griffon, and when the Air Supply Committee met Sir Henry Tizard on May 24 1941 it was agreed to once more place the Griffon Spitfire in abeyance. Not until autumn 1941, following reports of the Fw 190's low-level excellence, was the engine re-instated.

Alone in the wilds of Buckinghamshire, Martin-Baker were enterprisingly persevering with three fighter designs and with flight trials as distant as ever. By May 31 1941 their cost was £58,000, the company estimating a current loss of £38,000. The slow progress appalled the Ministry which regretted awarding a contract to a company motivated by the type of enterprise and unconventionality with which the Establishment finds difficulty in coping.

Wide-ranging discussion of the Spitfire's programme took place on June 16 1941. High-altitude Merlin 46/47 prototypes were expected in September, also two experimental Griffon prototypes, production of which could, if worthwhile, commence in March 1942. Proposed armament was six cannon or 12 machine-guns, fitting of which would surely have been difficult. The Universal Wing was also to be introduced bearing two 60-round belt-fed 20 mm cannon and four machine-guns, after the 520th Mk Vb, in September 1941. Previously the aircraft were armed with two drum-fed cannon and four machine-guns. Westland fitted the new feed on their 100th Mk V in spring 1942, and the Castle Bromwich works, on their 800th Mk V, from October 1941. Metal ailerons were to be applied to the 90th *et seq* Supermarine-built Mk Vs and would, from August 9 1941, be featured by all production Mk Vs. By June 1941 the two-stage Merlin 61 was being flight-tested and the Spitfire 'B' Type airframe being much redesigned to accommodate it. The MAP was concerned about delay, but not until November 1941 did the CRD agree to its trials in the existing Type 'B' airframe.

Hurricane tank-buster design, commenced at Esher on May 30, was coming along well. An alternative 40 mm Rolls-Royce cannon was being installed in a Beaufighter. Deep concern at the paucity of the tank-buster order was expressed by the CAS who feared these aircraft might be the only effective defence against invading tanks. His worry was placated by the ACAST whose concern was more

*The earlier two-speed supercharged Merlins incorporated a gear change to increase the revs of the supercharger impeller at high altitudes, whereas the two-stage engine had a second supercharger which fed the air/fuel mix by way of an intercooler.

with the Hurricane's ever-increasing all-up weight for, with two 44-gal drop tanks, the cannon-armed Mk IIc's 8,040 lb compared alarmingly with the typical production Mk I's 6,218 lb. Yet its take-off run remained only 310 yds. Eventually, 4,711 Mk IIcs were built.

June's major event was the delivery of the first production Typhoon. Two more followed in July and in succeeding months 6, 12, 0, 7 and 11. By the end of June Chadderton's first Tornado (powered by a Vulture V, first flown in the prototype on April 1 1941) was supposedly a few weeks from flying. June 9 brought a DTD's request for a Typhoon incorporating an exhaust-driven turbo-supercharger, discussed with Hawker on June 18 and sanctioned along with a 4-ft wingspan increase on June 30. The MAP asked on July 8 that eight .50 in machine-guns be made possible for the Typhoon's Universal Wing, all part of the improved Typhoon mentioned by Lord Beaverbrook. The aircraft was ever unlucky for, on July 11, *P5216*'s port undercarriage leg collapsed bringing more general annoyance.

Boscombe Down, where the Typhoon Ib had arrived for firing trials on June 22, accepted *R7576*, the first production Typhoon Ia, on June 3, at which time tropicalisation was being considered. Agreement of August 13 led to redesign of the aircraft's cabin, better rudder balance and improved finish. A Wright Cyclone engine was considered but soon forgotten in the belief that a better Sabre must surely be ready soon. *R7936*, the first Avro-built Tornado,

Comparison of the performance of the Tornado and the Typhoon.

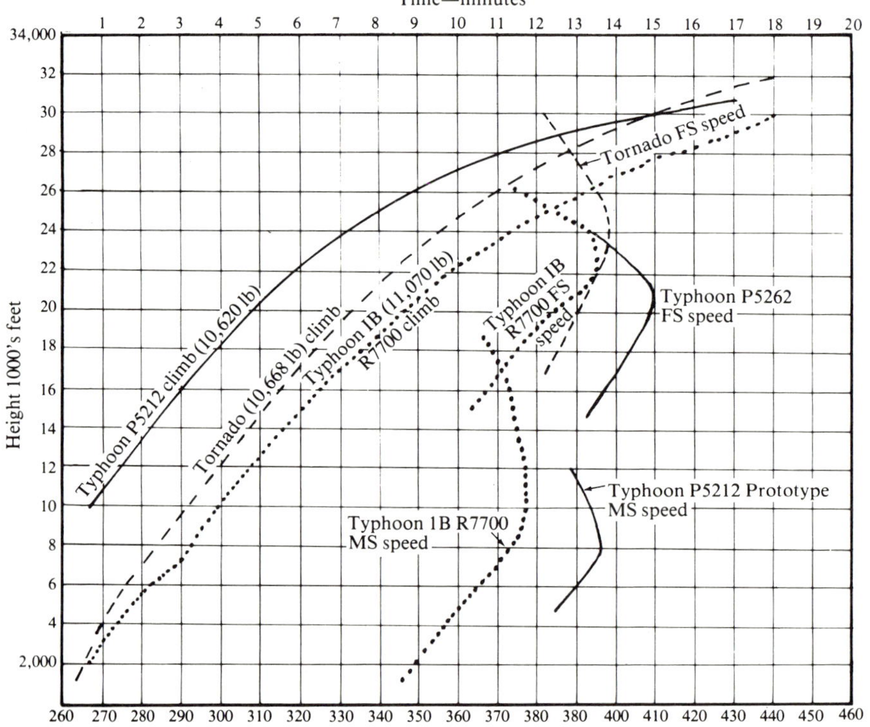

Top *Not until spring 1942 did cannon Typhoon Ibs enter service. P5216, the second Typhoon prototype, was the first to be fitted with cannon* (BAe, via Bruce Robertson). **Above** *Avro-built Tornado R7936 was used for Rotol and de Havilland contra-prop tests. It features the additional strengthening applied to Typhoons around the rear fuselage* (BAe). **Below** *An example of the RAF's first fighter-bomber, a Hurricane IIb carrying two 250 lb bombs. BE485 joined 402 Squadron in October 1941 and was written off on April 17 1942* (IWM CH4567).

Top *Typhoon Ia R7580 arrived at AFDU Duxford on August 31 1941, joined 56 Squadron on September 19 and was damaged in a flying accident on February 8 1942. Later used by 59 OTU (PRO).* **Above** *The fourth production Typhoon Ia, R7579, used in the preparation of the Pilots' Notes for the type. Early Typhoons like this example had dark green and dark earth camouflage and Sky under surfaces. Spinner was black (RAF Museum P9348).*

eventually flew on August 30 and their second was at Yeadon* for assembly instruction when the first one flew to Langley on October 15 1941 for trials.

Meanwhile, 2 (Bomber) Group's hazardous anti-shipping campaign was bringing ever-mounting losses so in August 1941 the Air Staff decided that Hurricane IIbs, each carrying 2×250 lb bombs, would replace the Blenheims, probably in November, by which time they might be carrying 2×500 lb bombs apiece. The 'Hurribomber' was clearly going to become a most useful machine.

Apart from Hawker's, the only new type of fighter was still Martin-Baker's MB 3. Slow progress — reviewed on August 13 — was attributed to novel features. There was belief that the Sabre would improve when in reality its problems mounted, alarmingly so for it was the only available 2,000 hp engine for high-performance aircraft — including Camm's Typhoon redesign.

The cannon story was exactly the opposite. Production was good and cannon were bringing fine success rates. From August 1941 .303 in Browning machine-gun production, running at 15,000 units monthly, was ordered to no longer increase although it continued until 1945. Eventual production totals were: 1938

*The intention was for Yeadon to be the second Albemarle production source, but choice switched to Brough to free Yeadon for Tornado, initially 200 aircraft, and easy to increase.

— 9,671; 1939 — 22,387; 1940 — 55,353; 1941 — 129,944; 1942 — 167,820; 1943 — 71,820; 1944 — 51,082 and 1945 — 3,704. Hispano 20 mm gun production amounted to about 2,300 guns monthly in summer 1941, the main proportion coming from BMAR. Yearly production amounted to 432 in 1939, 3,360 in 1940 and 12,022 in 1941 — all backed by spares amounting to about 200 guns per month.

Without doubt September 1941 marked a major turning point in the fighter saga. Fw 190s were in action over France and the Sabre was in very deep trouble. There can be no doubt that the Typhoon was prematurely placed in service, its failings being heaped upon No 56 Squadron which, on September 11 1941, was declared as having 're-formed' with four Typhoons. 'Plagued' might have been more accurate.

As the area around Duxford was drenched by the raucous Sabre scream, Sydney Camm was completing plans launched two months previously for the Typhoon Mk II. Faster, more reliable and having enhanced ceiling, it would be powered by a new version of the dreaded Sabre. Camm was asked on September 4 1941 for performance estimates for his new design. Rapid official appraisal resulted in the response that 'if the figures are anything approaching the firm's estimate the changeover is worthwhile and we shouldn't delay'. The CRD wanted to order two immediately, to which Hawker replied that they had an even better version in view. Comparative performance forecasts for the three new Typhoon variants were:

	Typhoon I (Sabre II)	Typhoon II (Sabre NS8M/IV)	Typhoon (NS11SM turbo blown)
All-up weight, lb	10,800	11,300	11,800
Maximum speed, mph/TAS/ft	409/21,000	455/26,000	440/35,000
Operational ceiling, ft	30,700	35,300	41,000
Service ceiling, ft	33,950	37,900	43,200
Endurance, hrs	1.54	1.73	1.7
Wing span, ft	41	44	46
Wing area, sq ft	274	295	322

Napier's claim of 2,010 hp at 26,000 ft for the Sabre IV seemed optimistic, but the Typhoon II was likely to gain these increases: from power 29 mph, ejector exhausts 4 mph, thin wing 9 mph, improved cooling 4 mph, total 46 mph. On the basis of these figures Hawker were, on September 17, informed of an intended order for two prototype Typhoon IIs featuring increased-span eliptical wings holding four cannon, and a Sabre IV forward-set to permit increased fuel load. Wing radiators in the Bf 109F-style would replace the Typhoon I's chin type. The ACAST expressed disappointment that six cannon could not be carried, before it was decided, on September 20, to reduce the wingspan to 42 ft, increasing relative strength. Permanent tankage was set at 180 gal allowing a 150-mile range. Jettisonable fuel tanks afforded a further 150 miles.

Plans to increase the Hurricane's punch were proceeding. *Z2326*, the Vickers 40 mm cannon-armed Hurricane IId tank-buster, first flew on September 18 and next day flew to Boscombe Down where its trials proved almost trouble-free. In October it demonstrated its potential by destroying a Valentine tank, following which the original order for 100 Mk IIds — already increased to 500

Top *Spitfire I P9565 carries a 30 gal fixed external fuel tank. This was an early attempt to extend the Spitfire's range for a bomber-escort role.* **Above** *A production Hurricane IIc Z3775: JX-Y of No 1 Squadron* (Flight International). **Below** *Hurricane IVd KZ193 during a maker's test flight* (RAF Museum).

in June — dramatically rose to 1,000 by March 1942. Another Hurricane II, *Z3156*, was on trial in an attempt to improve high-altitude qualities, but attempts to fit larger under-wing tanks to the Mk II resulted in excessive vibration. Relatively short range operations with long-range tanks were feasible, as tests with Mk IIc *Z3888* proved. With two 90-gal drop tanks, all-up weight 9,060 lb, the IIc still managed to get away from a hard surface in 310 yd. Without drop tanks (all-up weight 7,260 lb) its top speed was 328 mph at 22,000 ft compared with a clean IIb's 330 mph at 20,800 ft after a 9 min climb to height.

Great concern was expressed over the Sabre situation. Without the engine the future of the fighter force was in great jeopardy. The Sabre II was proving so unreliable that the MAP cancelled plans for two Sabre Production Groups. These would now build Bristol Hercules engines for bombers. Napier, with 1,000 employees, were producing a mere six or eight Sabres weekly. Lord Beaverbrook, considering the whole Sabre programme a 'fiasco', said forcefully that what Napier needed was 're-organisation, not more factories'. After a month in service Typhoon engines were plagued by hosts of problems. Shafts, driving oil and coolant pumps suffered torsion failures, with 6 out of 27 engines already afflicted. Cold starts were very difficult and there was so much concern over carbon dioxide fumes entering the cockpit that pilots constantly wore oxygen masks. On October 6 1941 Napier was instructed to check every Duxford Typhoon.

Any idea of going for the unorthodox Vulture had long passed. It had suffered persistent connecting rod fractures and the Vulture's performance in the Manchester had been atrocious. Little wonder that on October 15 1941 its development was halted to allow Rolls-Royce at last to concentrate on the Merlin 60 and Griffon.* Neither could rescue the Typhoon. The Director of Technical Development said that in the emergency facing them it might be necessary to have a radial-engined fighter, and he had already mooted a Taurus-engined Whirlwind. He must have been joking! The only real alternative to the Sabre was the powerful Bristol Centaurus — still in its infancy, delayed by the success of other Bristol engines. The MAP thought this idea worth trying and very soon a Centaurus reached Langley, running in Typhoon *HG641* on October 11 and flown on October 24 and 26 1941. How this could change the fighter scene was uncertain because it was far too powerful for the Typhoon I. The RAF reckoned a typical Typhoon Ia had a top speed of 412 mph at 21,000 ft — slower than Beaverbrook claimed — and was inferior in climb and ceiling to a number of other fighters. Not surprisingly, interest raced to its refined Mk II successor. Sabre-powered though, could it be very much better?

Cannon-armed Typhoons were long in coming, due to the slowness with which the Air Staff requested them, and because of the craze for the 'Universal Wing'. Additionally, there was no universal wish for cannon Typhoons, the C-in-C Fighter Command pointing out that different gun combinations suited different combat situations. For fighter-versus-fighter combat some pilots preferred machine-guns, obtaining quick deflection shots by using them. No 11

---

*The Vulture II gave 1,540 bhp at 15,000 ft. Thus fitted, the 10,600 lb Tornado reached 380 mph at 16,000 ft and 350 mph at 30,000 ft and had a service ceiling of 33,000 ft. The Vulture III was forecast to give 1,790 bhp at 20,000 ft and the Tornado a top speed of 413 mph at 22,000 ft and a 34,000 ft service ceiling. The Typhoon reached 410 mph at 22,000 ft and 33,500 ft. Fitted with a Centaurus it would attain 393 mph at 18,000 ft and have a service ceiling of 32,500 ft.

Fitting a Centaurus to Typhoon HG641 *produced an over-powered aeroplane. Undercarriage covering plates needed to be repositioned to avoid the exhaust pipes* (RAF Museum P6951).

Group wanted mixed armament, whereas squadrons outside enemy fighter range preferred cannon. Fighter Command was eager that Typhoon IIs should have a universal layout agreeing on October 9 to four 20 mm cannon, or two cannon and four machine-guns, either .303 in or 0.5 in — the latter to attract American interest. A contract for a trials wing came on October 13, Hawker extending the alternatives to include smoke layers, two 40 mm cannon and bombs.

Lightly loaded, the Typhoon prototype reached 422 mph in October 1941 compared with a Spitfire V's usual 378 mph, which showed the Typhoon's potential — when its engine was in good form.

A contract for two Mk II Typhoons (*HM595* and *HM599*) was placed on November 18 1941, with production to commence in spring 1943. Hawker promised a prototype would fly in March 1942. To hasten things, the MAP suggested using the fuselage of *P5212*, but instead a new pair was built, one to be held incomplete. With persistent optimism, Major Bulman reckoned a Sabre IV would be available for type testing by the end of 1941. Sydney Camm was far more realistic and on November 7 pointed out that the Typhoon II might have to rely on the Sabre II. Such was the case when production began in November 1943, late indeed.

Handling trials of the Typhoon I at AFDU Duxford, conducted at heights up to 20,000 ft, showed it good for turns and combat manoeuvres yet suffering severe vibration above + 4 G. That needed remedying. In an attempt to increase the ceiling, *P5216* was, on November 7, cleared for testing with extended wing tips. On the same day, but for the opposite end of the flight envelope, Camm submitted details of a ground attack Typhoon Ib able to carry two 44-gal or two 90-gal drop tanks. Clearly he had the foresight to realise the Typhoon's limitations might make it a prime attack fighter. Simultaneously he was able to confirm that the Hurricane could carry two 500 lb bombs.

If something worthwhile was to arise from the Typhoon's troubles, what was to overtake the Spitfire was truly remarkable for early flight trials of the two-stage Merlin 61 had gone extremely well. Although demands for a Spitfire thus powered were far less strident than for the Typhoon II, so good were the reports of the new engine that the CRD recommended the splendid new combination be accepted, long before a Merlin 61 completed a type test. Available information suggested an output of 1,560 bhp at 12,000 ft, and 1,370 bhp at 24,000 ft, conferring a radical advance upon the Spitfire's performance. First production Merlin 61s were expected in May 1942, indications being that a Merlin 61 Spitfire would perform well at 25,000 ft. Supermarine had full particulars of the engine when called upon to redesign the Spitfire to take it, recommending first fitting it into Type 'B' airframes so that all Mk Vcs could be converted and entry to service of what was now the Mk IX be speeded. This the MAP agreed to and the CRD informed the firm that Mk V production should be phased out in favour of the MK IX, the prototype of which was *N3297*, the original Mk III first flown after being re-engined with a Merlin 61 on September 20 1941, from the Rolls-Royce airfield at Hucknall. The Mk VIII would be the Spitfire airframe completely revised for the Merlin 61.

Boscombe Down trials of Spitfire *N3297* (Merlin 61) armed with two cannon and four .303 in guns and carrying 120 gal of fuel, showed its handling was good. There was relief at the absence of carburation and ignition troubles at great height, that engine cooling and intercooling were satisfactory. Noticeable gyro effect from the four-bladed propeller little changed the landing run. Radiator shutter positioning was important and when closed the aircraft reached an astonishing 42,500 ft — 10,000 ft more than the Typhoon. *N3297*'s retractable tailwheel and flush, bullet-proof windscreen would be featured only by the Mk VIII. Its loaded weight was 7,600 lb and power/weight ratio at 30,000 ft was 7.3 lb/hp against the Mk V's 9 lb/hp. Its only drawback was the none too impressive climb performance between 17,000 and 25,000 ft, as a result of which boost was increased from +9 lb to +12 lb for emergency use. Comparison made in November 1941 between the RAF's three latest fighters showed the 'hybrid' Spitfire IX fastest, achieving 421 mph at 27,500 ft, the Typhoon 412 mph at 21,000 ft and 381 mph at 23,500 ft attained by a Merlin 46 Spitfire.

So impressed was the ACAST that, on November 12 1941, he wanted to cancel all but Merlin 61 Spitfires. It was proposed that Spitfire V production, now running at about 110 aircraft monthly, be altered to give about ten Mk IXs a month by mid-1942, whilst letting the Vs run on a little longer at Castle Bromwich. Some evidence of the Merlin 61's excellence was evident when, during an intended 114-hr test run, the engine did 111 hrs before an auxiliary drive failed. On December 3 1941 Merlin 61 production was laid down as likely to be: 8 in February 1942 and in successive months 16, 40, 60, 60, 71, 80, 90,100, 110 and, in December, 100.

Martin-Baker, meanwhile, continued work on the MB 3, its time scale so slipped that on November 2 1941 N.E. Rowe stated that there was no possibility of it entering production irrespective of its merits. They seemed unimpressive and likely to show it inferior to the Typhoon and possessing a longer landing run, due to high wing loading. On December 24 the firm was told, 'It is not possible for the Ministry to place an order before flight trials have been carried out', to soften the blow to the enterprising. Martin-Baker responded by claiming that the aircraft was 'well advanced'. They could not deny it was overdue.

The Air Staff were more impressed by the Hurricane IId (tare weight 5,585 lb; loaded 7,656 lb), delivery of which began in December 1941. Twenty-five were scheduled to be built by May 1942, by which time delivery rate would be 12 monthly. Likely range of the Typhoon Mk IId was 400 miles, its angle of attack 10°. Belt feed would deliver 15 rpg, 1,000 guns being on order. The first dozen aircraft, armed with two 40 mm cannon and two .303 in guns, were earmarked for the Middle East.

December 1941 found the Air Staff drafting Specification F.10/41 covering the Sabre IV/Typhoon II whose airframe strength, judging by the Typhoon, would need particular attention and be able to withstand at least 550 mph EAS at 10,000 ft in a dive — which speed took the aircraft into compressibility effects. Fear of these around the radiator's fairing lips led to a RAE investigation. The chin-type of the Mk I was inadequate for a more powerful engine so an enlarged nose radiator, wing radiator aft of the rear spar or wing leading edge radiators were suggested. Least drag was produced by the third arrangement, shock stall occurring around 420 mph on the second layout and at 470 mph on the others. By mid-December 1941 Hawker had abandoned the idea of wing radiators, they were too complicated. This assumed the Sabre's trouble cured when intricate mechanical and metallurgical problems remained. Suggestions that Rolls-Royce take over Napier to sort things out were rapidly squashed; the staff would never have accepted that!

On December 18 1941 data from official tests of the Centaurus CE 4SM Typhoon, tested at 10,600 lb, became available and were unexpectedly impressive. Its top speed was 430 mph at 24,000 ft, economic cruise 239 mph at 20,000 ft and service ceiling 34,500 ft. It took 10 mins to reach 20,000 ft, estimated combat range being 550 miles. Take off over 50 ft took 790 yds, landing 980 yds. But production was impossible since the engines needed much development.

Two significant events occurred before 1941 passed. The new Fw 190 was showing a fine low-level performance, so Rolls-Royce were ordered to develop rapidly the single-stage Griffon III/IV for possible fitment in a Spitfire Vc airframe, this becoming the Spitfire XII whose prototype *DP845*, first flew on November 27. On December 23 1941 an Air Staff meeting proposed 100 be built on the Castle Bromwich line beginning in September 1942. A much modified Spitfire IV with two-speed two-stage Griffon 61 and designated Mk XX would follow, entering production in July 1943. The CAS wanted the switch sooner, but the CRD was alarmed that a low-level interceptor was going ahead when combat was taking place at ever-increasing height. He pressed for only six Spitfire XIIs prior to the Mk XX — and then only for trials. The meeting refused to agree to that, arguing that ample Griffon experience was needed. Since the 'B' airframe was involved, the Mk XIIs could easily revert to being Merlin aircraft. Also at this meeting it was agreed that work on the Spitfire floatplane be resumed and completed by May 1942 against rumours of an 'export' order for 11. The aircraft also had attraction as an island defender.

December 1941's other decision was to arm RAF fighters with air-to-ground rocket projectiles. Reports of effective use of such weapons against tanks and bomber formations from a British Military Mission to the USSR caused the ACAST to ask the CRD to investigate the value of rockets against ships and tanks. Rocket-firing trials commenced in November 1941. Necessary were ejectors for firing and an adaptation of the 3-in anti-aircraft rocket projectiles used on 'Z' projectors. Between November 1941 and spring 1942, RAE designed and developed suitable rocket launchers, tested on Hurricane II *Z2415*.

Throughout January 1942 Hawker concentrated upon the Typhoon II, its future brightening when, on January 1, Sir Henry Tizard suggested applying the Centaurus to make the aircraft useful as a ground-attack fighter. Centaurus production could reach 300-400 engines monthly, although only at the expense of desperately needed Hercules engines. Nevertheless, on January 24 design of a Centaurus Typhoon II was sanctioned, a hefty machine needing a stronger undercarriage than other Typhoons. Hawker was told to proceed with six development aircraft, as well as jigging and tooling for the absorption of 50 Centaurus engines a month. Sydney Camm was informed of this decision on February 3 and told that development was urgent.

The many failings of the Typhoon Is in service had prompted the sudden move. Carbon monoxide poisoning had caused a fatal accident on November 1 1941 following which Sabre Engine Modification 112 of January 23 introduced lengthened exhaust manifold stubs. This was ordered to be applied immediately, causing the grounding of all Typhoons. Various cures had been tried, including sealing possible cabin leak points, re-arrangement of ventilation systems and moving the exhaust pipe positions. Low-drag flame damping ejector exhausts were retained. The current system embraced six ejector pipes each fed by two branch pipes attached by flanges to common exhaust ports for the top and bottom cylinders. The proposal was that this should be retained, but that exhaust outlets should be 4 in further outward from the engine centre line to use existing stocks of pipes, to each of which a 4-in extension was welded. All operational training aircraft were so modified — 18 at Duxford, one at Wittering — as well as experimental aircraft. Then the level of cockpit contamination became acceptable. Recent dive tests proved satisfactory, although speed reduction needed watching.

For five months Typhoons had given troubled service at Duxford where many problems were discussed on February 10 1942. Fuel tanks emptied unequally, bringing excessive trim troubles. Reinforced tyres and tubes were needed, following tube bursts due to wheel brakes overheating. The pilot's seat needed improved springing and Duxford's rough surface caused broken tailwheel brackets. There were too many loose rivets in the airframes and double images in the gunsights. Viscosity valves in the oil-cooler had been troublesome and pressurised tanks preventing fuel from boiling had led to oversight of the vulnerability of fuel pipes near the cockpit floor. The Perspex fairing behind the pilot and a rear view mirror tested on *R7595* were major advances applied to most of 56 Squadron's aircraft, the clear rear canopy section being introduced on the 164th *et seq* Typhoon built. None of these changes could cure the affliction of the Sabre — and could it still be improved for the Typhoon II?

That design encountered further misfortune — the Universal Wing. So bad was the bickering over armament that it seriously delayed the design. In addition, during F.10/41's design stage, the Air Ministry reversed a decision; it must now carry six cannon. It was impossible to fit them into the thin wing and incredibly, to the end of 1942, arguments over armament raged. Additionally, the fitting of .50 in machine-guns also posed a peculiar problem. They required large ports in wing leading edges and lengthy blast tubes because these guns needed to be mounted well back in the wings on sufficiently sturdy pick-up points. Such holes might reduce the aircraft's speed. Also necessary would be complex heating systems, as well as large access doors suitable for cannon or machine-guns. Because of these complications C-in-C Fighter Command agreed in October 1942 to forgo the Universal Wing and accept only four cannon, because of the aircraft's promised high performance. Could that be achieved?

The cannon-armed Typhoon Ib entered production in February 1942. Camm had first suggested such armament in August 1938 and in July 1939 the go-ahead was given for an experimental pair of wings with four 20 mm cannon. Further delay arose when the CAS stated that production could only start when cannon had proved effective in the Beaufighter. Not until October 1940 did the Air Staff sanction production of cannon-armed F.18/37s and it was April 1942 before Typhoon Ib delivery commenced at a slow rate.

Alongside the Typhoon, Hawker continued to develop the Hurricane. Indeed, on February 24 1942, two days after a Mk IIb first flew carrying two 500 lb bombs, the Air Staff agreed (reluctantly) to accept 1,250 Hurricanes in 1942-43 in lieu of a theoretical loss of 843 Typhoons for which engines were unlikely to be available. Camm had, on January 16 1942, explained to the DTD that aerodynamic clean-up of the Hurricane II would give it 20 mph more speed, as a result of which more Hurricanes were ordered on January 23. Three days later the MAP increased the Canadian order too. Production of the Mk IId was proceeding well, although there was concern that so little armour protected the pilot.

With eight UP rockets being easily carried by the Hurricane Hawker faced renewed demand for a Universal Wing including fittings embracing two 40 mm cannon, drop tanks, rockets and bombs. This was featured by the Hurricane Mk IV first flown on March 14 1943. The RAF was to receive 524 Mk IVs, and the first use of rocket-firing Hurricanes came on September 2 1943. March 17 1942 found the MAP and the Air Ministry reviewing Hurricane production. Gloster would build 1,780 Hurricanes and 371 Typhoons by 1944, a contract raising Hurricane production from 4,788 to 6,238 being signed on March 25 as weekly output reached 77 aircraft. Eventually, 2,952 Hurricanes were diverted to the USSR. The extended life and value of the Hurricane was quite unexpected.

Although basically unchanged, mid-war Hurricanes exhibited many improving features. HW603 *shown here has cannon and bombing equipment* (BAe).

As Hawker veered towards ground support fighters, Supermarine concentrated on interceptors. Higher speed, ceiling and climb rates were attained by Spitfire IXs than by the refined and better armoured, tropicalised Mk VIII whose ceiling was cut by 500 ft due to revised engine cooling. Its top speed during service trials was 416 mph at 27,500 ft. Rolls-Royce, fearing this might prompt more calls for Griffon production, stressed that the ideal two-stage Griffon Mk 61 could not be mass-produced before August 1943. Supermarine, on the other hand, sensing pressure for Griffon Spitfires would increase, devised one around the Mk IIb engine despite reiterated instructions that such a variant was unacceptable. Discovery of this caused the CRD to tell Supermarine, in no uncertain tones, that it was 'directly contrary to our principles'. More was behind that outburst than the company knew for, in late April 1942, a Griffon IIb Typhoon II seemed a worthwhile alternative to a Sabre version. By entering production mid-1943 it would be a useful low-attack fighter-bomber/tank buster, a logical Hurricane successor.

As an interim measure Hawker was authorised, on April 14 1942, to adapt the Typhoon Ib to carry two 500 lb bombs on wing strong points intended for 90-gal drop tanks. They would be useful features for the Mk II whose 186 gal permanent tankage showed a 40 per cent increase over the Typhoon I's. Design work on the Hawker P.1016, a Griffon 61 Typhoon II, was also going ahead. To bring the Hurricane into line it needed provisioning for two 90-gal drop tanks, giving it in loaded state a still air range of 1,550 miles.

Success of the interim Spitfire IX prompted Sir Wilfrid Freeman on April 28 1942 to remind Supermarine that the Mk VIII must be in full scale production by the end of 1942, and that Griffon 61 Spitfires must be available by June 1943 to replace existing fighters — apart from Hurricane IIds — by early 1944. Reliance would then switch to the turbo-blown or three-stage Sabre Typhoon II and a two-speed two-stage Centaurus ground-attack aircraft based upon the Hawker Tornado. The Griffon Typhoon II was seemingly a transient idea. Freeman at this time stated that he had every confidence in the Sabre becoming satisfactory, but his belief was not widely shared. Attempts to persuade Major Halford to leave de Havilland to concentrate upon improving the Sabre were not successful. Instead, a Deputy Managing Director was appointed to revise production planning at Napier for, by mid-May 1942, insufficient Sabres were available even to meet Typhoon I production. Nevertheless, Freeman's confidence in the Sabre II remained. He believed that all would be well in June and that the Sabre IV was as suitable for development as the Centaurus. The Deputy Controller General tactfully pointed to 'a considerable body of opinion in industry' which saw the Sabre remaining a doubtful engine, diagnosing this as 'due to the inadequacies of Napier's design staff'.

Although the Centaurus and Griffon could replace the Sabre, Rolls had on offer only the one-stage two-speed Griffon II. With + 12 lb boost it gave 1,720 bhp at 7,500 ft and 1,490 bhp at 14,000 ft. In desperation, due to the critical Sabre situation, production of the Griffon II was suddenly ordered. Supermarine's hunch would pay off! Two months too late came the type testing of the two-stage Merlin 65 offering 2,300 hp at 500 ft and 2,060 hp at 15,750 ft. Its development was out of phase with any fighter programme, otherwise it might have supplanted the Griffon.

As soon as Hawker discovered Griffon production was sanctioned they applied it to the Typhoon II, in both Mk II and 61 forms. Camm presented his

Few Brewster Buffaloes reached Britain, where they were unsuitable for front line service. Many operated in the Far East and were mercilessly dealt with by the Japanese. W8243 is depicted.

proposals to N.E. Rowe on April 10 and was told that Rolls-Royce would build the Griffon 61 as a complete power unit. Camm's estimates suggested speeds of 370 mph at 9,000 ft and 400 mph at 22,000 ft for a Griffon II Typhoon II, and 419 mph at 19,500 ft or 430 mph at 31,500 ft using the Mk 61. The latter version would attain 30,000 ft in 13.6 mins and possess an operating ceiling of 40,000 ft against the Griffon II's 31,500 ft. It was immediately pointed out to Camm that almost all Griffons were earmarked for Spitfires, (which Supermarine had yet to learn) and that no Griffon 61 even for the Spitfire XXI would now be available before September 1943. Opposed to any engine change for the Typhoon I, the CRD, on May 12, told Hawker to make a Griffon trial installation in the second Typhoon II.

Wing construction for the first Typhoon II was proceeding slowly, its first flight slipping towards September 1942 and precluding production before July 1943. Revised armament and radiator positioning further induced delay. When talk of a Griffon added problems, to boost the design's fortune the DTD ordered six prototypes on June 17, two each of Sabre, Griffon and Centaurus versions.

Spitfire IX deliveries were underway, coincidental with ever more disturbing reports of Fw 190 capabilities. Although slower than the Typhoon II, it appeared able to out-climb it to achieve its best performance between 14,000 ft and 20,000 ft. In the region for which the Spitfire IX was intended the Fw 190 was inferior, so the new mark was entering service as tactics were changing.

April 1942 had seen the Hurricane IId enter service at Shandur, in the Western Desert. On June 7 No 6 Squadron flew the first Mk IId operational sorties, from Gambut. The following day the first two attacks were delivered costing two of the six aircraft. Tanks were found awkwardly dispersed among other MT nullifying the 40 mm guns' potency, although on June 15 seven tanks were claimed destroyed. In June 43 sorties were flown, 35 in July, but a hasty withdrawal to the east reduced operational flying.

Since 1940 American fighters had been coming into RAF hands. One type was the Curtiss Hawk H-75C bought by France, diverted to Britain and there renamed Mohawk I. Its Wright Cyclone GR-1820-G205A gave a maximum output of 1,200 hp at 4,200 ft. The aircraft had a top speed of only 302 mph at 14,000 ft, reached 15,000 ft in 6.2 mins and had a best rate of climb of 2,600 fpm at 8,000 ft. Diverted from Norway was the Mohawk II (Pratt and Whitney Twin Wasp R-1830-SC3G, highest output 1,050 hp at 7,700 ft) which, when

Above *Curtiss Mohawks also served mainly in the Far East. Although over 50 were shipped to Britain, fewer than ten were active in the country.* **Below** *A Curtiss Hawk 81-A Tomahawk I, AX900. Tomahawks commenced arriving in Britain late 1940. AX900 underwent A&AEE trials between February and June 1941 and successively served with 41 OTU, 1472 Flight, 168 Squadron and 1681 Flight. It was written off on May 11 1944 (Westland Aircraft).*

tested at 5,962 lb, reached 300 mph at 10,000 ft and reached 15,000 ft in 7.3 mins, after taking off in 290 yds. Most of the 215 Mohawks supplied to Britain were Mk IV (Wright Cyclone GR-1820-G205A). Pending supply of Hurricanes they were used by No 5 Squadron (December 1941 to June 1943) and 155 Squadron (August 1942 to January 1944) for bomber escort and ground attack operations in Burma. This version featured four .303 in Browning guns, two in the fuselage and two in the wings. Although manoeuvrable, the Mohawk was no glittering performer.

In October 1940 the first ex-French Curtiss Hawk H81A-1 Tomahawk 1 (1,040 hp Allison V-1710-33) reached Britain. A second Tomahawk batch, Hawk 81A-2s and 3s, arrived late in 1940, their two .50 in fuselage guns being supplemented by four .300 in wing guns. These Tomahawk IIs had an all-up weight of 7,270 lb. Their top speed was around 330 mph at 15,000 ft, a fuel load of 84 gal allowing 2.8 hours' duration when cruising at 20,000 ft — better than that of contemporary British fighters. Overload fuel tankage allowed 132 gal, increasing the weight to 7,646 lb.

The Tomahawk was America's best fighter and, by December 1940, 1,041 were on order for the RAF. The British intention was to employ them in North Africa, but shortage of shipping space delayed this. Eventually they gave quite good service in that war theatre. Only 112 Tomahawk Mk Is, 61 Mk IIs and 43 Mk IIbs (the latter with US equipment) were retained in Britain. Four Mk Is joined 234 Squadron in December 1940 for operational assessment. Poor medium altitude performance, mixed calibre guns and tricky flying qualities brought rapid Fighter Command rejection. Army Co-operation Command had been begging for Hurricanes to replace vulnerable Lysanders and on January 1 1941 the Air Staff stated that because Hurricanes were scarce, the Tomahawk should be tried in the tactical reconnaissance role. This is the 'latest American fighter, with a speed of 350 mph', the AOC was told. That was an exageration — the best a Mk I had clocked was 338 mph at 16,000 ft and it took 7.5 mins to reach 15,000 ft; it would be vulnerable in combat. Army Co-operation Command had no choice but to take what was offered and, in February 1941, No 26 Squadron became the first to fly Tomahawks. Eventually there were 15 squadrons; Mk IIs being introduced in August.

Morale in the Command was then low. To raise his own morale, Flying Officer G.C.H. Jackson of 400 Squadron, making a training flight, crossed, un-authorised, to Cap Gris Nez. Unfortunately, during his low level venture a wing tip of *AH812* graced the ground. He had something to explain after landing back at Odiham!

Air Marshal Sir Arthur S. Barratt, AOC-in-C Army Co-operation Command, managed on August 14 to persuade Fighter Command to agree to Tomahawks flying *Rhubarbs* from its stations, after. 303 in guns for mixed type ammunition were installed in selected aircraft. The first authorised *Rhubarb* was flown by a 239 Squadron Tomahawk to the Ostend area on September 19 1941, such operations also by 26 and 400 Squadrons continuing until spring 1942.

June 1941 brought along the Bell Airacobra, a streamlined form devised around a 37 mm cannon firing through the propeller hub and placed on the centre line to reduce recoil effect. RAF Airacobras carried a less exotic 20 mm

Of the 675 Bell Airacobras shipped to Britain only 49 are known to have been active there. RAF examples had a 20 mm nose cannon, and a few had extensive Sky finish as shown here (RAF Museum P6161).

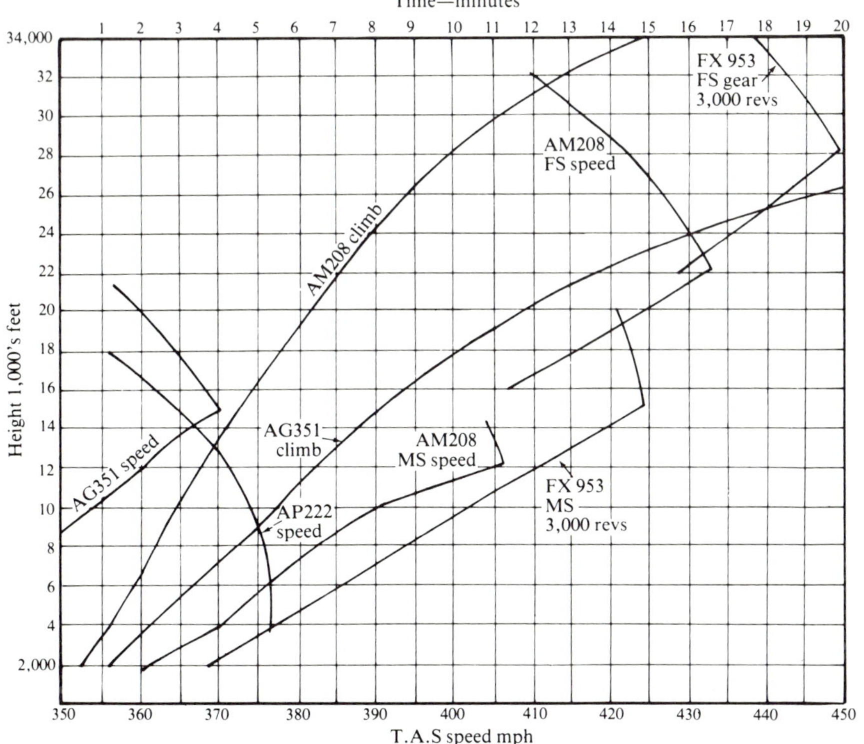

Graph of the Mustang's performance at full load. (AG351 = Mk I, Allison V-1610-F3R, at 8,443 lb; AP222 = Mk I, Allison V-1610-F4R; AM208 = Mk X, Merlin 65 experimental, at 8,650 lb; FX953 = Mk III, Packard Merlin V-1650-3. At 2,700 revs, maximum speed: MS 406 mph, 20,600 feet; FS 438 mph, 33,000 feet.)

cannon, plus two .50 in guns in the nose and four .303 in wing guns. The low-altitude rated 1,150 hp Allison V-1710-E4 was sited amidships aft of the pilot to ease cg problems. This meant a troublesome 10 ft transmission shaft, and a nosewheel undercarriage added more unconventionality. Britain ordered 675 Airacobras in 1940, accepting that the low altitude-rated engine rendered them only suitable for low attack operations. Typical operating weight of the Aira-cobra was 7,840 lb, its service ceiling 29,000 ft; best rate of climb 2,040 fpm at 10,300 ft; time to 20,000 ft 11.7 mins and top speed 335 mph at 13,000 ft. An improved Allison raised the speed to 365 mph at 15,600 ft and a few Airacobras with .50 in wing guns carried the designation Mk Ia.

Its flying qualities satisfactory, the Airacobra was placed in 601 Squadron in August 1941. On September 27 the squadron prepared to operate from Manston while alternative uses as a Turbinlite satellite, bomber or convoy escort aircraft were reviewed. On October 6 1941 No 601 Squadron flew three sorties against ground targets near Dunkirk. Another was flown the following morning before three Airacobras joined 615 Squadron's Hurricanes during a shipping sweep. Poor serviceability was encountered and the aircraft withdrew from operations. There was, also, another reason. When the nose guns fired their electronics set up strong magnetic fields causing astonishing compass deviations claimed as high as 165°! What was needed was a wing-mounted Pioneer Distant Reading

Compass and 20 were promptly ordered in the USA. According to the VCAS the compass was 'not as bad as rumoured', but he admitted that it was 'uncontrollable'.

On November 8 Duxford entertained some Russians, after which only 49 of the 675 Airacobras on order were assembled here; the rest were donated to their maker — or the Soviet Union! American fighters had proven useless so far, but this was soon to alter. After buying Harvard trainers from the North American company, the British encouraged the firm — not eager to produce P-40s — to design, in April 1940, a fighter to meet their need. Within days a proposal for the sleek NA-73 was to hand. An order for 320 was placed on May 29 with the proviso that the prototype must fly in the autumn. American engines were far inferior to the Merlin and it was unfortunate that the beautifully clean lines of the NA-73 were coupled to the Allison V-1710. The machine featured the unusual, high-speed thin laminar flow-wing whose maximum thickness well aft across the chord allowed smooth air-flow over a major part of the wing surface, although slow speed performance was impaired. Much attention was also devoted to tucking the radiator well into the rear fuselage, again to reduce drag.

Keeping to the deal, North American flew the aircraft on October 26 1940. Before it crashed during its fifth flight as a result of fuel starvation, 620 NA-73 Mustang Is were on order, the first flying on April 16 1941, proving very rapid development was possible. It had reached 394 mph at 15,000 ft although it was

Below *The beautifully shaped North American Mustang, America's best wartime fighter. AG345 illustrated prior to painting was the first example intended for the RAF* (North American). **Bottom** *The fourth production Mustang I, AG348* (North American).

Above *Angular wing and tail shapes of the Mustang led to fears that it might be mistaken easily for a Messerschmitt Bf 109, hence the yellow wing bands for identity* (RAF Museum 5939-10). **Above right** *Elongated nose, twin radiator baths and four-bladed propeller characterised the superb Spitfire IX, of which BS289 was a production example* (Rolls-Royce).

without operational equipment. The first examples were packed for Britain in September 1941. No 26 Squadron, on January 5 1942, was the first to try a Mustang. The Air Ministry was so impressed with its unexpected performance that, on January 24 1942, ten Army Co-operation Command squadrons were ordered to be equipped, for low level fighter reconnaissance tasks. Unfortunately the Allison V-1710 F3R 12-cylinder vee liquid-cooled engine peaked at 12,000 ft, giving 1,150 hp when running at 3,000 rpm. After discovering how smoothly the engine ran, on low revs, its low fuel consumption was found to allow flights of an amazing 4.1 hrs cruising at 15,000 ft, equivalent to a still air range of 990 miles and no mean achievement. The armament of two .50 in nose guns and two more in the wings was supplemented by four .303 in guns, and the 11 ft 10 in track was ideal for rough field operations.* The instrument layout was good but the cockpit became very warm — and again there was compass trouble. Tare, the Mustang I weighed 6,288 lb and loaded, carrying 140 gal of fuel, weighed 8,625 lb. Its top speed, radiator shutter closed, was 370 mph at 15,000 ft. To reach 16,000 ft took 8.65 mins, to 28,000 ft 24.6 mins. Service ceiling was 30,000 ft.

Fitted with the Allison V-1710 F21R (supercharger gear ratio reduced to 7.48:1 from 8.8:1) the Mustang I attained 377.5 mph at 4,000 ft, but the service ceiling fell to 25,000 ft. One Mustang, *AM107,* was tested carrying two 40 mm 'S' cannon, its top speed being 341 mph at 10,000 ft. The Mustang II, similar to the Mk I but with an F4R engine, weighed 8,200 lb loaded and had a speed of 401 mph at 4,400 ft, and 409 mph at 10,000 ft when the V-1710 F20R was installed. Such high performance figures had been unexpected. Farnborough investigations in June 1942 attributed them to a drag level of 50 lb/100 ft per

*Browning Mk II .50 in machine-guns began arriving late 1942, 6,554 having been delivered to Britain by the end of that year. Main deliveries came in 1944; 35,764 guns in all being delivered.

sec, compared with 65.5 lb/100 ft per sec of the average Spitfire. This, it was reckoned, was largely due to the use of a laminar flow wing which was forecast to add only 8 mph to the Typhoon II's speed.

AFDU Duxford found their Mustang I faster than the Spitfire V to 25,000 ft, peaking to 375-380 mph at 15,000 ft, compared with the Spitfire's 340 mph. Climb, however, was inferior, falling to 1,000 fpm at 25,000 ft and at all altitudes the Spitfire out-turned the Mustang. For long-range low-level operations, though, the Mustang was an excellent aeroplane. On May 5 1942 a Mustang I of 26 Squadron, Gatwick, flew the first Mustang operational flight and attacked targets near Berck-sur-Mer. By July 1942 Mustangs were regularly flying such sorties. Their ample range was demonstrated on November 27 when they first penetrated into Germany. Only the Allison engine restricted the Mustang's performance, so could a Merlin be wedded to the airframe, preferably a Merlin 61? Rolls-Royce estimates of April 1942 suggested a likely top speed of 441 mph at 25,600 ft. Packard Motors were already building Merlins in the USA and were preparing to produce the Mk 61. The idea was put before the Air Ministry and on June 9 Air Marshal F.J. Linnell informed the Americans that the Air Staff and the Ministries were keen on the idea. Indeed, the VCAS wanted six Mustangs immediately converted but that would mean six much-needed Spitfire IXs would be lost. Instead, three Mustang conversions were ordered on June 15.

Spitfire IXbs were now in service with 64 Squadron, Hornchurch. Their first uneventful operation was flown on July 28 1942. Two days later, when covering 'Hurribombers' of 174 Squadron attacking St Omer airfield, a fierce battle developed and the new Spitfires claimed four Fw 190s. No 72 and 402 Squadrons were equipped with Mk IXs in July, No 133 and 401 in August. No 11 Group then rotated the aircraft among its changing squadrons.

As the Mk IXs (the original order for which was increased because the Mk VIII was not ready) were reaching Hornchurch, the Allies were switching hard to the offensive and the new Spitfires became useful escorts for USAAF B-17s. Implicit, too, was the need for massive air support for ground forces. Rolls-Royce therefore adjusted the balance of the Merlin 61, enabling its use at medium altitudes at the expense of some superiority at height, and had adjusted the supercharger in June 1942 causing the engine to peak between 10,000 and 20,000 ft. This led to the 1943 Merlin 66.

Cabinet demands for action on the Sabre crisis were very strong by mid-1942. A survey showed that flying hours per engine failure between December 1941 and May 1942 were 340, compared with the Merlin's 500 hours during its first six months of service, and 1,000 hours currently. An MAP estimate of June 2 showed a shortfall of 1,400 Sabres by late 1943 whilst production of 2,500 Typhoon Is and 760 Mk IIs was likely. What could be done to utilise these airframes? Switching to the Centaurus remained the only possibility, supplementing it with some Griffons. Yet the Centaurus was, as already noted,unsuitable for the Typhoon I. Urgent redesign of the Mk II was ordered with a view to building 700 Centaurus examples, thereby releasing Sabre IIs for Typhoon Is. That many Centaurus engines meant supplementing intended Accrington production and thereby losing precious Hercules engines. On June 16 the MAP proposed instead that engineless Typhoons be stored awaiting Sabre IIs.

A Centaurus Typhoon II, suddenly the favoured variant, seemed likely to be fast. But, without major engine modifications, it would have a low ceiling and an unimpressive climb rate. Pressure was therefore applied to hasten the Sabre IV to mid-1943 availability and to increase Sabre II output. Without Typhoon IIs the RAF would need to rely upon the later Spitfires — or perhaps turn to the Merlin Mustang? Official forecasts suggested a Merlin Mustang would reach 427 mph at 25,500 ft against Rolls-Royce's claim of 441 mph. Certainly it would be fast, have long range and a high ceiling. Agreement was reached between Britain and America on July 20 1942 for 1,200 Merlin Mustangs to be built between January and October 1943; 50 per cent going to Britain in exchange for diversion to the USAAF of an equivalent number of Packard Merlins. Doubt existed over Packard's ability to have the Merlin 61 in production by January 1943 and, additionally, there was fear that if the aircraft became really successful the Americans would want all the Mustangs! Proposals for a Packard V-1650-3 (Merlin 61) USAAF Mustang (XP-78) discussed on August 9 1942 resulted in an order for three. By then the Typhoon was operational.

Although a few scrambles had taken place, the first Typhoon Wing operation came on June 20 1942 when 56 and 266 Squadrons mounted a diversionary sweep as Bostons attacked Le Havre power station. July 19 found the third squadron participating for the first time when 609 Squadron helped to support Bostons making a low level attack. Not until August 9 did an enemy aircraft fall to Typhoons, a Ju 88 shot down by two pilots of 266 Squadron off Yarmouth. And still the Sabre was in deep trouble. Its realibility was subjected to analysis at Duxford on July 7 when Group Captain J. Grandy said that new problems 'had a disturbing effect, since between May 10 and July 2 1942 modified engines had been installed in squadron aircraft, and still 35 engine failures had occurred'. Twenty-four resulted from sleeve valve defects not previously encountered. The only alteration which could have caused them was associated with slow running section settings resulting in excess fuel washing away lubrication for piston and sleeve valves, allowing excessive wear particularly during slow running periods. All engines were now adjusted to a weaker setting. Any running on over-rich mixture remained liable to cause failure, but despite the problems Typhoon squadrons remained operational. Five days later a report attributed piston ring failure, etc, as due to excessive oil breathing and high oil

Above *Under-wing bomb racks changed the Typhoon into an excellent aircraft whose fighter-bombing and rocket attacks played a large part in the success of the Normandy Campaign in 1944. Shown is one of the first fighter-bombers, R8831:EL-U of 181 Squadron (RAF Museum P2691).* **Below** *Hybrid Typhoons were common by 1945; the example shown here retains car door entry to its revised canopy, yet has a four-bladed propeller (BAe).*

consumption, along with occasional holes in the pistons. All engines were re-inspected and two aircraft carried out intensive flying using slow running, weak mixture fuel. Better news of the Typhoon, given later in July, was that it would safely carry two 43 gal long range tanks, two 500 lb or two 1,000 lb bombs on wing stations.

Fighter Command, finding the 'Hurribomber' very effective, logically pressed for Typhoon fighter-bombers. Despite Sabre problems the Air Staff acquiesced and on July 21 sanctioned formation of two squadrons in September 1942 to conduct Army support operations. In August three Hurricanes fitted with rockets completed effective tests. Rocket rails for Hurricanes entered production in December 1942 and were released for service in January 1943.

Various roles for the Typhoon were being considered. *R7881* had AI Mk VI radar, but a night-fighter Typhoon would need longer patrol duration, airbrakes and a clearer windscreen set closer to the pilot. It was incompatible with the Havoc and thus unsuitable as a Turbinlite satellite. In a personal aside the AMSO said, on August 6, 'If the Typhoon never comes up to the required standard it can be passed to the Russians' — before he discovered that its engine was utterly unsuitable for cold weather operations!

How well, though, did a typical Typhoon now perform? At the end of August 1942 tests of Mk Ib *R7700* (Sabre II) driving a 14-ft three-bladed propeller

Above MN173, *a Typhoon Ib (Sabre II), has the ultimate clear view canopy. Cannon have been removed because the aircraft was under test at Luton with Napier* (Napier). **Below right** DP845, *the first Griffon-engined Spitfire, originally the Mk IV and developed into the Mk XII* (Supermarine, via Bruce Robertson).

showed top speeds of 376 mph at 8,500 ft in MS gear and 394 mph at 20,200 ft in FS gear. In 6.2 mins it reached 15,000 ft and 25,000 ft in 12.4 mins. Service ceiling was 32,200 ft; the fastest rate of climb being 2,000 fpm at 17,800 ft in FS gear after take off at 11,070 lb. *R7646* used for bombing trials weighed 8,663 lb tare. Fully loaded, carrying 154 gal of fuel and two 500 lb bombs, it weighed 12,184 lb. Go-ahead for 300 sets of equipment to convert Typhoons into fighter-bombers was authorised on August 18 1942, the first modified aircraft joining 181 Squadron at Duxford on September 7.

Sabre suitability for more production accepted, the MAP ordered increased capacity. From Napier's new London factory at Earl's Court Exhibition Centre under Park Royal's control, monthly output would be 125 engines. Monthly schedules were now for 410 Sabres, sufficient if the Sabre Firebrand fighter was dropped and 100 Centaurus engines became available monthly for the Typhoon II. Alas, all did not flow smoothly, the new London works proving a fiasco. Alterations to those premises were halted and they were switched to other uses.

The Typhoon II was now so much removed from the Mk I that it was renamed 'Tempest' on August 6 1942. The Tempest I would have a Sabre IV, the Mk II a Centaurus, the Mk III a Griffon II and the Mk IV a Griffon 61. A final decision taken on August 25 was to site the Mk I's radiator in the wing leading edge, after a few fuselages had been completed with chin radiators. Drag reduction and improved rear protection had provoked the change, but placing the oil cooler in the wing would reduce fuel space allowing only 123 gal, much less than required. Therefore, the oil-cooler was positioned beneath the nose allowing 140 gal of fuel to be carried internally. Already there were suggestions for night intruder, fighter-bomber, low attack and rocket-firing anti-tank Tempests.

Summer 1942 saw plans agreed for the Spitfire XX/P1 (alias Mk IV/P1) the prototype for which, *DP851*, first flew on August 8. Production examples would initially have Griffon IIb engines, later machines the Griffon 61 forecast to give them a speed of 410 mph at 23,000 ft, climb to 32,000 ft in 15 mins and a ceiling of 37,000 ft. For the first time a Spitfire design included a fighter-bomber role allowing for one 500 lb bomb, a 30 gal drop tank or tankage giving a 1,000-mile range. The detailed specification issued on May 26 1942 called for Mk XXs to be built at Castle Bromwich and by Philips & Powis, South Marston. All eventually emerged as sophisticated Mk 21s.

By August 1942 Spitfire airframe production was out-stripping Merlin 61 output, although the prototype Mk VIII *JF274* first flew in November and production deliveries began in April 1943. There was then a shortage of 100 engines for Mk IXs. Therefore, early VIIIs had Merlin 70 high-altitude engines complemented by pointed wing tips. Roughly based upon the Spitfire Vc, the Mk VIII had internal tankage for 124 gal, the additional fuel being in the wing roots. It had stiffer yet smaller ailerons, and a retractable tail wheel. Concern over the Fw 190 brought a decision to instal low-altitude rated Griffon IIIs in Mk Vc airframes, then designated Spitfire Mk XII, 50 of which were ordered to be completed by the end of the year, another 50 by April 1943 — which proved very optimistic.

Merlin Mustang interest ever increased. A telegram despatched on August 27 1942 to the British Commission in Washington instructed it to get Mustang production expanded to allow 120 engineless airframes to be sent monthly to Britain and, if possible, another 200 to the Middle East, India and Australia. In return the USAAF in Britain would receive an equivalent number of Spitfire IXs. Doubting the viability of that aircraft in 1943, the Vice-Chief of the Air Staff even suggested ordering 3,000 Merlin Mustangs, half for the RAF. But engine production never allowed this to materialise. Instead, the RAF helped out the USAAF in Britain by handing over some Spitfire Vbs.

August 1942 brought to a climax the MB 3 programme. Late March had been the contract completion date, but it was repeatedly pushed back. On March 26 the firm claimed the aircraft six to eight weeks from flight testing. Wing cannon were in place, but neither their feed nor the undercarriage oleos. Meanwhile, the firm repeatedly wrote to the CAS extolling the virtues of their late aircraft. Uncertain of the truth he asked for others' opinions. The general concensus was that at low levels it would be fast, but that its ceiling would be poor. In view of the proximity of completion, the prototype was allowed to proceed with June 30, then July 31 1942, as latest acceptable delivery dates. Hold-ups with the radiator meant that it was late August when *R2492* vacated Denham for Wing and flight trials. The maiden flight took place on August 31, the Gallay radiator being immediately found to be unsuitable. Engine temperature rose to 135°F

Excessively aggressive in appearance, the ill-fated Martin Baker MB 3 at Wing in August 1942 (IWM MH5015).

compelling Captain Baker to land after only one circuit of the airfield. He maintained that too much taxying had brought about the trouble. Napier representatives agreed, stating that there was no danger in a further flight. Once the coolant temperature was low a late evening flight was made, but again the coolant became excessively hot, boiling after one circuit of the airfield. As a result the aircraft was grounded. Captain Baker stated that it had handled well, particularly during the more alarming moments and a heavy landing which proved the undercarriage to be strong. It was hoped that in 48 hrs radiator troubles could be cured.

Some problems were overcome when, in the hands of Captain H.V. Baker, the MB 3 crashed on the evening of September 12 1942 during its tenth flight and killed its pilot. Engine overheating had not been solved. Official enquiries showed that the project had greatly exceeded the contract price based on cost plus five per cent profit to a maximum of £25,500. The exact expenditure was not possible to disentangle from the firm's other work which the Ministry claimed had, for Martin-Baker, been most profitable.

Following the accident Martin-Baker improved their fighter design, shifting the radiators from below the wings and into the rear fuselage. As a result January 31 1943 was set as delivery date for the second prototype but it was June 1944 before it flew. One must surely praise the persistence of the firm. In 1942 Captain Baker tried hard to interest North American and General Motors representatives in Britain in producing the fighter, but after detailed analysis they turned it down. They had, already, a winner on the way, estimating their XP-78 Merlin Mustang likely to have a speed of 445 mph at 28,000 ft and a service ceiling of 42,000 ft, figures which the MB 3 could never attain. It would need a lightened or extremely powerful version to achieve something like that.

Chapter 7

Improving all species

Fighters were becoming heavy due to weighty armour, armament and hefty power plants, a trend which needed to halt before manoeuvrability suffered unacceptably. A turning point came on August 13 1942 when a new piston-engined fighter with outstanding qualities was first seriously considered. Limited power and unreliability of gas turbines meant that one more reciprocating-engined fighter was needed prior to the jets taking over, and it would have to be outstanding. Emphasis would rest upon climb and manoeuvrability, even to the detriment of speed. A precedent for a similar situation being found in the 1980s was being established. This new fighter would need to climb at 4,500 fpm to 20,000 ft, have a top speed of 450 mph and have the range essential for Far East operations. Designers' comments upon these ideas were invited, while the new commitment was not to be allowed to impede work already in hand. Production would feature as much as aerodynamic considerations, and hopefully the entire programme could be lodged with one contractor. A radical concept ahead of its time was the possible formation of a consortium whose sole preoccupation would be with the fighter. Firms already working at peak capacity were thus being eliminated, although Hawker were likely to want to break from the Typhoon line, while Boulton Paul, eager to step away from their unlucky Defiant days, were both possible firms to handle this major project.

The requirement was for a high-speed, single-seat fighter superior in climb, speed and fighting qualities to any known enemy machine and operating at medium altitude. Forthcoming jets would handle high-level combat. Optimum performance at 20,000 ft was coupled to a multi-role capability. High rate of climb was linked to rapid acceleration, fast roll and essential weight saving. Permanent tankage would allow for 10 mins' maximum climb, 15 mins' fast chase or combat at 20,000 ft and 30 mins for cruise return. Auxiliary tankage would equal 40 per cent of permanent load. Four 20 mm wing cannon would have sufficient ammunition for 15 secs' firing, bombs being alternatives to extra wing tanks. This was needed in the F.6/42.

The emphasis laid upon a new design source (an idea with which the Air Staff were much smitten because they considered the two main teams merely developed existing success), was rapidly reviewed by Sir Roy Fedden. Mr Ratcliffe of Folland Aircraft was his favourite for design team leadership. Under H.P. Folland, Ratcliffe had worked out the successful Gloster fighters, and that had impressed Sir Roy. On August 14 he hurried to discuss his ideas

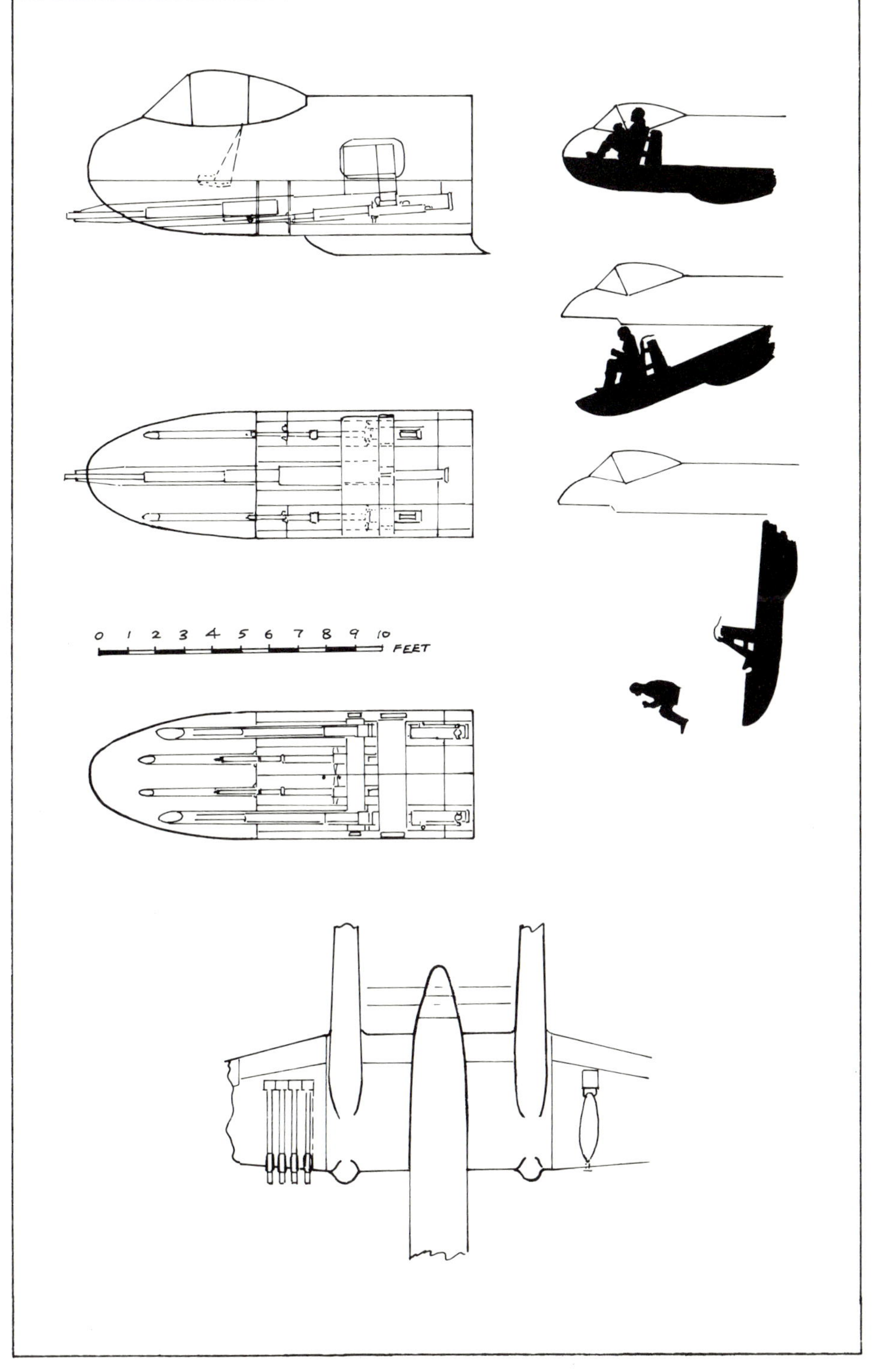
0 1 2 3 4 5 6 7 8 9 10 FEET

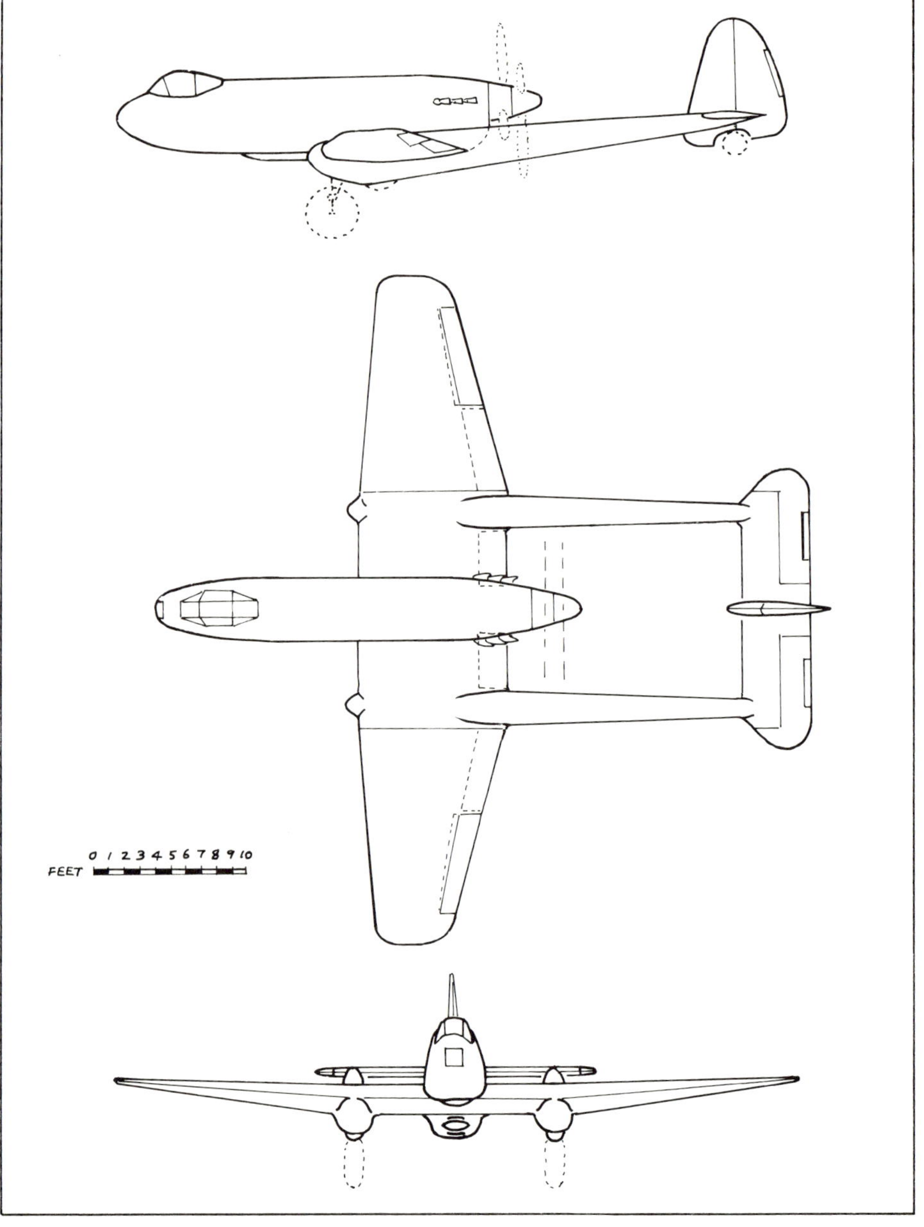

This page and opposite *The Boulton Paul Series P99 low-attack aircraft, a single-seater powered by a Rolls-Royce Griffon single-stage engine.*
Installation details of 47 mm gun flanked by two 20 mm cannon are shown, and below an arrangement of two 20 mm cannon and two 40 mm cannon. Alternative armament was three 40 mm cannon. A most ingenious form of emergency escape for the pilot was provided in both the P99 and P100. The top view shows the normal flying position, the second after 0.2 sec and the third after 0.4 sec, after which the pilot then used his parachute conventionally.

with Ratcliffe, the two concluding that the fighter could be rapidly designed. But there was a snag in that Folland was not listed as likely to be capable of performing the task. With a design staff of only 25 and 20 trainees they would need to drop their E.28/40 and recruit more staff, which would not be easy. On hearing of this notion the MAP rapidly claimed that, although Ratcliffe might be able, Folland was too small and lacked any production facilities.

N.E. Rowe bypassed the whole matter, seeking views of the more conventional firms — Westland, Boulton Paul, Hawker and Vickers — particularly enquiring about the value of contra-rotating propellers and the laminar wings which had benefited the Mustang. He was told that the former could much delay the aircraft whereas laminar flow wings — did they not result in very long landing runs? Engine choice was perforce limited to the Griffon 61, Centaurus CSM (Mod II) and Sabre NS 43SM. Airspeed was already planning a high-speed day fighter and the first, surprising, submission, came from A.E. Hagg's team. Combining his design with a low attack fighter, he considered was not easy, but the importance of acquiring the contract made the attempt worthwhile. Official needs were for five prototype and eight production aircraft flying by December 1944, 176 examples by July 1945, 1,185 by the end of 1945 and no less than 3,935 by June 1946 — if the war continued that long. No wonder many firms were interested.

Boulton Paul claimed that existing engines would not provide the necessary performance, reckoning the most promising layout would include a Sabre engine in a tail-first design. Although its structure would be light, the MAP's, view was that it seemed still to be carrying 800 lb too much weight, so a few days later the company changed its mind submitting, on September 8 1942, plans for an orthodox, single propeller aircraft barely meeting the needs, along with Sabre or Centaurus versions featuring contra-props. The MAP considered all likely to be too slow, but a contra-prop tail-first version would have the speed needed. Boulton Paul were trying for a wing loading of 40 lb/sq ft to compete with the Fw 190 and to better its tight turning circle. The Sabre tail-first version (all-up weight 9,892 lb) was reckoned able to reach 446 mph at 20,000 ft, a conventional tractor Griffon machine 431 mph at 20,000 ft, its Centaurus variant 414 mph at 20,000 ft and the Sabre form 437 mph at 20,000 ft. Like the Miles and Airspeed ideas, all were thought too slow and rejected.

Westland, on September 3 1942, submitted plans of a Griffon 61 fighter, top speed 440 mph at 25,000 ft and ability to climb at 4,050 fpm at 20,000 ft, take 8.75 mins to reach 30,000 ft and have a 40,000 ft service ceiling. Vickers in Project RKP/63554 opted for a Centaurus-engined fighter claiming excellent climb rate. Folland, although uninvited, also submitted plans on September 3, claiming that their entry could replace obsolete aircraft still being built. Hamble works would construct 80 wings and fuselages monthly. With MAP help, and a collaborating 'shadow factory', production could reach 25 aircraft monthly. An interesting feature a Folland's design was an engine in the form of a complete 'power egg'. All designs featured five-bladed Rotol propellers. Although not suggested, a contra-prop would not have improved climb, but it could have aided manoeuvrability and reduced take-off swing.

In mid-September design comparison was undertaken — including Hawker's later submission. Westland's rather large machine had a conventional radiator slung beneath its power plant. The four-bladed 12 ft 6 in propeller would certainly have had a high tip speed. A Griffon had been chosen because it

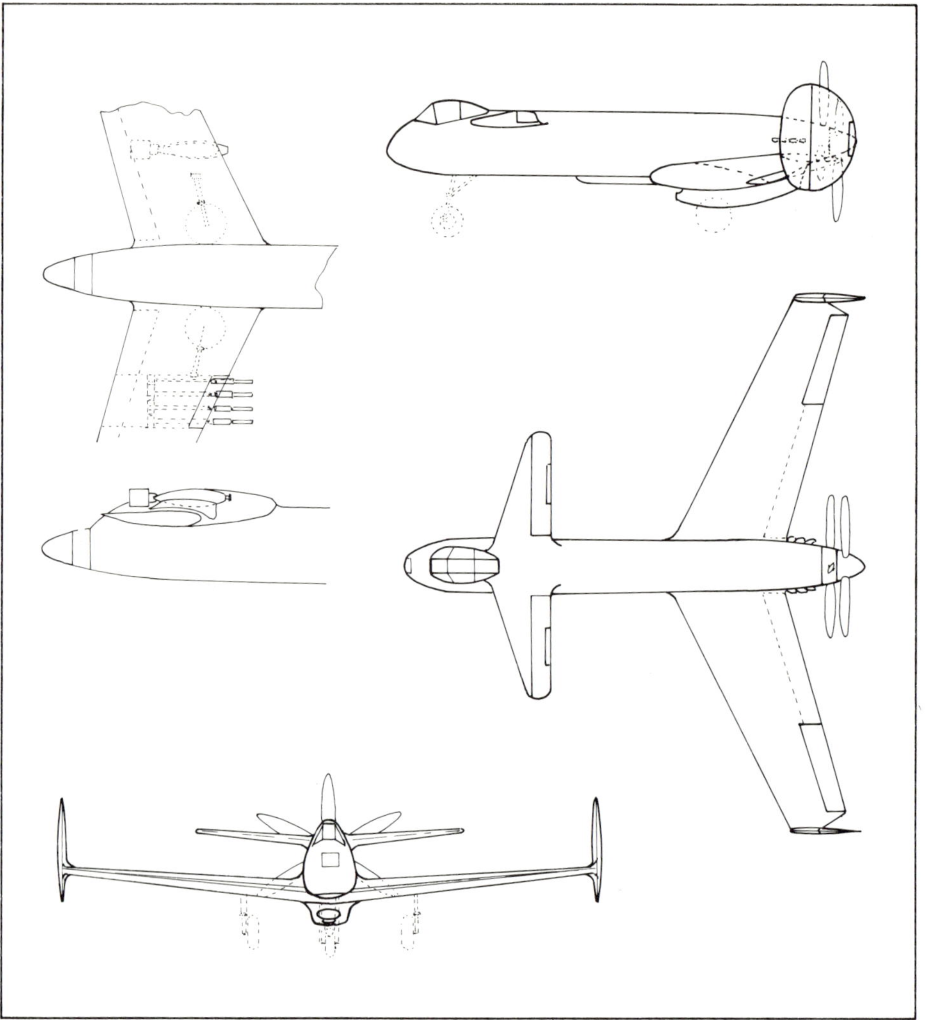

Boulton Paul P100 low-attack single-seat fighter (Rolls-Royce Griffon II). One of a family of low/medium-altitude fighter projects designed by Boulton Paul in 1942 around part of the overall requirements embodied in Specification F.6/42. All of the company's designs were considered to be 'too futuristic'. Indeed, new fighters of the 1980s utilise some features of the P100 including foreplanes and a clear forward view, here reminiscent of the Fairchild A-10 whose role cannot be all that far removed from that intended for the P100. A loaded weight of 13,450 lb was assumed, giving a wing loading of 40 lb/sq ft. Maximum sea-level speed was 298 mph, maximum speed at 17,000 ft was 335 mph. Normal rate of climb to 3,000 ft was 1,640 ft/min, take off to clear 50 ft, 700 yds, landing run, 680 yds, normal range, 795 miles at 265 mph at 7,000 ft. At the base of the nose could be fitted either four 20 mm cannon, two 40 mm and two 20 mm cannon or one 47 mm cannon and two 20 mm guns. Two 500 lb bombs or eight rockets could be wing mounted. Placing the engine aft raised problems for a pilot baling out. To avoid the counter-rotating propellers, the pilot left the aircraft in an emergency from the lower part of the aircraft, which dropped and was then hurled away, leaving the pilot to free fall, as from the P99.

seemed likely to be available on time. Supermarine's entrant was substantially a Spitfire 21 with a re-rated Griffon. Nothing special there. Airspeed's only novelty was an annular cowling radiator in the nose incorporating an engine cooling fan. Enterprising, its elongated propeller shaft and special fan drive seemed to afford no real advantage. Folland opted for a small propeller, a larger undercarriage than needed and also a scattered gun installation. Lastly came Hawker's three 36 ft-span fighters, P.1018 (Sabre IV), P.1019 (Griffon 61) and P.1020 (Centaurus IV) owing much to the Tempest. For the Sabre version they claimed 465 mph at 22,000 ft, the Centaurus type 450 mph at 22,000 ft. Were there any prospects of the likely needs of the late 1940s, maybe the 1950s, being met?

Griffon-engined designs weighed around 8,800 lb, Centaurus types about 9,800 lb and Sabre projects a little more. Centaurus designs appeared best with Folland's affording the clearest pilot view. Camm's submission was an improved Tempest II. Westland's latest version with a novel 'horse collar style' radiator around the Griffon had an air-cooled intercooler, rejected by Rolls-Royce. All Boulton Paul submissions were too futuristic and Supermarine's claim that the Spitfire 21 met the need was unacceptable. That mattered little to them for they hoped to build a laminar flow wing Spitfire.

The importance of the new fighter was such that the RAE was asked to investigate the merits of all designs. Success, the RAE stated, depended upon power/weight ratio, reckoning Hawker's design likely to encounter stability problems as well as having an untried carburation system and a Centaurus engine hard to armour against frontal attack. A contra-prop Sabre was distant and Napier's annular radiator involved a 1-ft-long untried propeller shaft. Giving the order to technically able Westland would inhibit Welkin development, for they were a small firm.

Eventually the RAE opted for Folland's design, although it was realised that the firm could never see it through. Secondly they placed Hawker's project, thirdly Airspeed's and lastly the Vickers and Westland ideas. Hawker's experience made them favourite to receive the contract — except with the RAF. Before any order was possible there was much to consider, not least the activity underway with a wide assortment of fighters. Although their workload was considerable, Hawker was a strong firm geared to large scale design, development and production programmes and allied to Gloster. The fighter-bomber Typhoon was being prepared for service, a requisition for 40 examples to be modified by Glosters for No 181 and 182 Squadrons being placed on September 17. Gloster produced and fitted three sets daily, commencing September 21, this rising to ten a week at Brockworth and eight at Staverton. From October Typhoons would leave the production lines with fighter-bomber provisioning. On the Prime Minister's instruction, three were to be demonstrated in the Middle East — when, and if, shipping space allowed. Mainly for that theatre, a dozen Vickers 'S' cannon were being built monthly. With 500 on order this was to rise to 50 a month, then hopefully 100. Success against various types of armoured vehicle had been achieved by the 'S' gun, resulting in a contract for 2,000 ground-attack Hurricanes placed on September 6, nearly all for the Middle East and Far East.

Troublesome Sabres were still being returned to the makers for repairs after 40 hrs running, due mainly to sleeve valve trouble. Difficulties with the Sabre I were declared unsolvable, all hope resting with the Sabre II expected soon.

Prototype Tempest V HM595 *retained many Typhoon features including tail unit and modified, clearer view canopy* (IWM MH4952).

Hawker encountered administrative bumbledom when it was discovered that no contract existed for an aeroplane called the Tempest! On September 5 cover for the Hawker Tempest I was amended calling for the 'Special Typhoon I/P1 Sabre 4'. Its permanent fuel needed to be sufficient for 5 mins take off power, 10 mins at maximum climb rate, 15 mins all-out at 20,000 ft, 30 mins for economic cruising at low altitude and auxiliary tankage 40 per cent of the permanent load.

After six months' delay the prototype Tempest, *HM595*, first flew on September 2 1942 and was powered by a Sabre II, the Mk IV not being ready. To avoid compressibility effects, wing leading edge radiators had been intended, but *HM595* had the Typhoon style. This alteration, failure to obtain castings and over-optimism at Hawker had all contributed to the delay. By the end of October the prototype had flown 20 hrs and exhibited directional instability. A shielded horn balance of 8 sq ft added to the Typhoon-type fin and rudder had no effect, so the fin area was increased by 2.7 sq ft, the rudder by 4 sq ft and then the stability became neutral. Serious longitudinal stability had also been encountered and tailplane buffeting around 370 mph ASI. Discussion of the aircraft's shock stall took place at the RAE during which the thickness/chord ratio of the wings was pointed out as, unusually, different from the tailplane and that the tail stalled first. The RAE tests had shown the superiority of wing leading edge radiators and in November Camm stated that the second Tempest would have them. By then it had been decided to increase tailplane area by 4 sq ft, although some problems remained because the wing was further back on the Tempest than on the Typhoon and it appeared there was nothing built in to compensate for this. To help speed the Tempest, Camm's request that it carry only cannon was agreed to because such guns were plentiful. BSA production had peaked, at 1,650 monthly. All production Tempests would also have sliding cockpit canopies.

The Tempest was not allowed to overshadow more immediately available

The fitting of rocket weapons to the Hurricane increased its potency, such aircraft serving mainly overseas. BP173 was used to test the system at A&AEE (RAF Museum P12036).

designs and on October 12 the first Typhoon fighter-bomber left Gloster for the RAF. Boscombe Down's trials of Hurricane IV *BP173/G* (tare weight 5,898 lb), the multi-role ground-attack version, were underway. With two 250 lb bombs aboard it weighed 7,822 lb loaded, 8,749 lb when carrying eight 60 lb RPs and 8,083 lb with 2 'S' cannon. Loaded with eight 60 lb RPs, its top speed was 261 mph at 12,000 ft.

While Hawker switched to aircraft for the attack role, Supermarine maintained their concentration on interceptor and escort Spitfires. The example attracting most attention now was not in their hands, but was a Mk Vb, *AA937*, at AFDU Duxford. Without permission the staff had clipped its wing tips, reducing wing area by 6 sq ft, then they plugged the tips with wooden fairings before trying it on October 2. There was an immediate improvement in the rate of roll, coupled with very light ailerons, bringing it more in line with the Fw 190. It also altered direction faster than a normal Spitfire, performance being best at low altitudes, although it remained fairly manoeuvrable up to 30,000 ft. At 35,000 ft, though, it was barely controllable. Ailerons remained very light even at high speed and might cause a pilot to be uncertain of his element of control. When flying low it was easy to droop a wing.

On October 14 AFDU's commander reported these findings to Commander-in-Chief, Fighter Command, telling him that the top speed at 10,000 ft was about 8 mph faster than the normal Mk Vb's, 5 mph faster at 20,000 ft and unchanged at 25,000 ft. Rate of climb was generally slower, it taking 6 secs longer to reach 20,000 ft. Take off took longer and landing speed rose by 5 mph. Set against this was unanimity that the clipped wing Spitfire V performed excellently at low levels and was well suited to face Fw 190 tip-and-run raiders. When news of this experiment reached Air Marshal R.S. Sorley, ACAST, he at once conveyed considerable anger to Fighter Command and more so to Duxford's commander. To the latter he wrote 'I am to point out that such a major alteration is entirely unauthorised and should not have been undertaken

A clever move undertaken at AFDU Duxford was the clipping of a Spitfire V's wing tips, as shown here on AA937:AF-O, one of their later Spitfire Vs (RAF Museum P12090).

within your Command. It is quite possible the aileron setting on the aircraft may be similar to the original Spitfire ailerons, which have proved quite dangerous. This aircraft is not to be flown again — except to Boscombe Down who will test it to see if it can be applied to the Mk XII — provided the aileron control is satisfactory.' Among his colleagues he expressed a different point of view, agreeing with Air Marshal F.J. Linnell that it was 'worth trying' at Boscombe Down. As for the daring souls of AFDU he stated that whilst he had 'rebuked them for the unorthodox alteration', he thought the idea might be most useful on the Mk XII, maybe the Mk IX, although he felt that performance above 25,000 ft must suffer seriously.

By the first week of November Boscombe was showing much approval of AFDU's initiative. Supermarine were ordered to make 20 sets of parts to allow 91 Squadron's Mk Vs to be modified. On November 17 Fighter Command bravely admitted that a Mk IX had already been clipped. Promptly another 19 sets of necessary parts were ordered, suitable for Mk Vcs, IXs and XIIs. The first of the latter emerged in October 1942 and during the A&AEE trials its top speed was found to be 396 mph at 18,600 ft. Clipping the wings had undoubtedly improved its low level handling.

Many in senior positions failed to understand just how radical and rapid were the changes taking place in tactics and weapons and that low-level operations were already of paramount importance. Apart from a few high-altitude raids in late summer 1942, low-level fighter-bomber attacks became increasingly annoying. Typhoons began to operate against them in September and later a good answer was the Griffon-powered Spitfire XII, but it was not a universally popular machine. Indeed, Air Marshal F. J. Linnell, in October 1942, told the Controller General that the Mk XII was 'only a stop gap', that he was strongly opposed to its production extension and that Spitfire development must surround the two-stage Merlin and Griffon 61. He felt so strongly that Supermarine had somehow defied officialdom that he was tempted to halt the

Tropicalised Spitfire Vb AB344 *fitted with a 90 gal slipper tank* (RAF Museum P5923).

Mk XII entirely. That would have been an error of judgment. Linnell, though, viewed the Spitfire as a medium and high altitude fighter, the Hurricane as the low flier. Indeed, after considering the low-rated Merlin 32, designed for the Navy's Barracuda, he authorised on December 28 1942 its trials in a Hurricane.

Problems of supplying sufficient fighters overseas had largely been overcome by direct deliveries of US aircraft supplemented by the despatch of Hurricanes by sea. Trials continued, however, to increase the Spitfire's duration. A Mk V with 90 gal external tank managed 337 mph at 19,500 ft and had a ceiling of 35,000 ft. After using, then dropping the tank the aircraft's still air range was 1,100 miles at 20,000 ft cruising at 170 mph IAS. On October 27 two Spitfires, each carrying a huge 170 gal tank, reached Malta after 5¼ hrs flying from Gibraltar. Both had 13 gal of oil and 43 and 47 gal of fuel left respectively. Their long-range tanks were not jettisoned and five similarly-fitted Spitfires followed.

A radical alternative to long-range tanks had been under examination since summer 1942, the towing of fighters by bombers. It was far from easy. Early tests involved a Wellington Ic towing a Hurricane locked upon a bridle. Flying at around 140 mph was difficult for the fighter and the Hurricane had also to fly offset from the Wellington's line of flight so that the fighter pilot could see the tug. Because of the bifurcated bridle, cannon armed Hurricanes could not be towed. Spitfires could only be towed on an assymetrical bridle because of an even worse view ahead. Late 1942 and early 1943 Spitfires *BF274* and *BF371* were used for towing trials by Flight Refuelling under Sir Alan Cobham's direction. Main overseas delivery would be of the Spitfire VIII and to provide fuel for its 8-hr flight it was suggested that it tow a Malinowski fuel trailer, possibly extending duration to 10 hrs and range to 2,300 miles. Take off at 10,350 lb was structurally feasible and a Wellington III was forecast, using a 700 ft tow, as able to tow a Spitfire 2,100 miles in 10½ hrs at 8,000 ft. Trials continued into 1943 and not until August 26 1943 were they halted, as 'not operationally viable'. Little known is the suggestion that Meteor jets could be towed overseas and that in June 1944 Liberators were considered as tugs for *Highball*

Mosquitoes needed in the Far East. An even more incredible suggestion was for an additional, jettisonable fuel-carrying aerofoil to be fitted forming a 'Mosquito biplane' with additional lift and range. Air Staff and Ministries have always taken a conservative view of somewhat avant garde ideas, fighter towing and such like proving no exception.

Greater interest was aroused by the Merlin Mustang. On October 8 the British Government tried to persuade the Americans to send a considerable quantity of Merlin Mustangs to Britain. The CRD maintained his belief that the aircraft's performance was being over-estimated and that its rate of climb would be inferior to the Spitfire IX's. It was now the only US fighter in which the RAF had any strong interest. Inhibited by having a small 10 ft 9 in diameter propeller, the prototype Mustang conversion flew at Hucknall in October 1942. A new flush undercowl, sealed radiator scoop and more intercooling could not prevent frequent engine cutting between 20,000 and 24,000 ft, due to fuel feed trouble. But flown against the Spitfire IX, with the same power available at 11,000 ft, the Mustang 'walked away' leaving the Spitfire unable to catch it. The prototype, *AL975,* had a top speed of 427 mph at 21,000 ft — only 8 mph below that forecast.

Elated by such success, the MAP representatives in Washington on November 5 now proposed 400 Mustangs be brought to Britain for rapid conversion. What they failed to mention was the non-availability of Merlin 61s. Rolls-Royce were sceptical of the whole scheme, believing that if large orders for the engines were not very soon placed the deal must fail — or only Packard engines be used. Rolls-Royce claimed that Merlin 61 delivery was retarded due to lack of fuel pumps. The MAP countered by acquiring American units — which needed modifications. Pressed to speed the entire programme Rolls-Royce, still sceptical, asked how this could be achieved without the airframes. The MAP concluded that even a good A&AEE report on the Merlin Mustang would fail to convince Rolls-Royce that large orders would follow. Rolls-Royce then laid out a conversion using the more plentiful Merlin 28, but the scheme soon fell by the wayside.

The position by November 11 was that *AL975* was undergoing performance tests with a larger propeller, *AM208* was soon to fly and *AM203* was complete. All were scheduled for delivery by December 5 for official trials, with *AL963* and *AM121* to follow before 1943.

Quite drastic events had unfolded at Napier and, in December 1942, control of the company passed to English Electric whose engine experience was limited but whose production know-how was good. Sir George Nelson became Managing Director of Napier, controlling activity at Acton and Liverpool and he immediately halted work on the Sabre IV. Three times unsuccessfully type-tested, its production application to the Tempest I now ceased, but its novel injection carburettor was an idea suitable for other Sabres. Nelson had the 50 Sabre IVs built, renamed Mk II(E) for psychological reasons and in an attempt to improve the Mk II's reputation. Such activities were unnecessary for the Rolls-Royce Griffon, though. December saw completion of the first Mk 61 which gave power outputs of 2,300 hp at 500 ft and 2,060 hp at 15,750 ft.

As if its Sabre troubles were not enough the Typhoon, which began fighter-bomber operations on November 28, had run into a very different problem. Throughout 1942 there had been increasing evidence of possible structural weakness. The aircraft vibrated badly, but that was not the main cause of inex-

Mustang AL975/G, the first to be fitted with a Rolls-Royce Merlin, arrived in Britain aboard the SS Swansea *on April 30 1942 and was used mainly by Rolls-Royce in development work leading to the Mustang III and P-51D* (RAF Museum P9025).

plicable accidents, in which the fuselage just ahead of the tailplane was prone to fracture. In an intensive modification programme by 13 MU Henlow, over 300 Typhoons, in four months, had strengthening plates fitted around the rear fuselage. Although these strengthened the aircraft they did not eradicate a problem eventually traced to elevator flutter induced by incorrect balance, in some instances aggravated by extended undercarriage fairing doors which suddenly caused high loading factors on the structure. Alleviation eventually came in 1944 with the fitting of the larger Tempest-type tailplane to the Typhoon.

December 1942 brought a résumé of fighter policy by the Air Staff who issued a statement pointing out that 'the fighter is an aggressive weapon. It has been of great use against enemy transport, and needs high speed manoeuvrability at low heights. The Hurricane has turned out to exceed what anyone ever thought would be its useful life. The Centaurus Tempest seems an obvious successor for low attack, but it is not coming very quickly. We cannot get better performance at low heights without the use of bigger engines. If we do not plan engine production we'll be in trouble.' Perhaps most important of all was the realisation that Fighter Command could now dictate the terms for battle using the height it prefered.

Much attention remained directed towards the F.6/42, the light, multi-role fighter. Folland of Cheltenham revealed a plan to build the airframe and have Bristol cope with the engine installation, leaving another firm to attend to armament. The MAP flatly rejected that scheme. Air Staff thinking was being directed now towards two F.6/42 variants, one having maximum performance below 20,000 ft and a second for higher regions, both with top speeds of 450 mph and excellent climb characteristics. The issued specification applied to the former and the Air Staff still favoured Folland's submission. The CRD in examining 1945's requirements eventually concluded that it was best and most economical to develop existing designs, which would not be done if the Folland was chosen.

An improved Spitfire was already underway. A low- to medium-altitude fighter with a Merlin RM14SM was reckoned likely to climb at 4,500 fpm at 17,000 ft and reach 430 mph at 20,000 ft, with the aid of contra-props. The conclusion was that a developed Spitfire was likely to be faster than the Folland but slower in climb. An RM15SM Spitfire could cope with higher level fighting, would climb at 2,000 fpm at 38,000 ft and have a speed of 450 mph at 35,000 ft, again with contra-props. Could, then, a revised Spitfire replace the F.6/42 scheme and remain viable until jet fighters were available?

The CRD also thought that the Tempest should be equally improved, likewise the Griffon Spitfire 21. Five days later, though, N. E. Rowe wrote to Boulton Paul, Airspeed, Vickers and Westland turning down their F.6/42 ideas. That left only Folland and Hawker to be dealt with.

Air Ministry officials were far from happy over suggestions for 're-hashing' old designs. Certainly an improved Merlin Spitfire would have the performance to take it well into 1944, but by then they held that its lack of protection from 20 mm fire, limited roll rate and restricted forward view would soon render it obsolescent. Control and manoeuvrability of the Folland might be better, pilot's view too. Air Staff attachment to air-cooled engines remained strong, indeed was strengthening because of their reduced vulnerability during low level strike sorties. The Air Ministry informed Sir Wilfrid Freeman that they could not agree to halting the Folland, but would agree to keeping the Spitfires in the front line to give Folland time to develop their machine.

Just what was being done to update the Spitfire? Summer 1942 brought a suggestion, following the Mustang's amazing performance, that a laminar flow wing suitable for the Spitfire 21 and Mk VIII should be developed, and would improve the Spitfire's rate of roll. Redesign would also remove the fear that engine development and power might outstrip the airframe limitations of the Spitfire. Work therefore commenced in October 1942. By mid-November Supermarine had to ease production and had designed a straight, tapered 35 ft-span wing, with an area of 210 sq ft, root chord of 100 in and tip chord of 40 in. Thickness/chord ratio would be 30 per cent of the inner 70 in of span, tapering to eight per cent at the wing tip. Larger ailerons were thought likely to need less balance. Most important, though, was the very smooth aerofoil section, the

Supermarine constantly improved the Spitfire, the most refined Merlin version being the Mk VIII typified by JF880's *increased rudder area and 'C' type armament.*

A line of hybrid Spitfire IXs, BR349 nearest in high-altitude fighter colours. Both this and its neighbour (MA399?) have extended wing tips and are believed to belong to 238 Squadron.

deepest part of which was well aft of the leading edge. This sharpened the latter, reduced drag and offset the compressibility effects increasingly encountered as speeds increased. In the process, gone was the familiar Spitfire elliptical wing. Into the Mk VIII's fuselage would be installed either a Merlin RM14 or RM15 giving an estimated climb rate of 4,100 fpm at 18,000 ft and a ceiling of around 48,000 ft on a loaded weight of 8,800 lb.

Farnborough's December investigation of the laminar refined, smooth-flow wing Spitfire suggested a top speed of 470 mph at 39,000 ft whereas Supermarine was claiming 500 mph. A critical Mach No of 0.72, occuring at 475 mph at 39,000 ft and shown in tunnel tests, might be bettered by aerodynamic refinement. Propeller disc limitation would be 11 ft — an important aspect, for it was now realised how important this could be for fast fighters driven by powerful engines. Might this much revised Spitfire become, virtually, the high-altitude part of F.6/42? Such was feasible, and would leave Hawker to continue as designers of low attack fighters.

Chapter 8

Radial return

Throughout 1942 Hawker had worked on a Centaurus-engined fighter following January 1942's prediction of 50 engines being available monthly in 1943. Indeed, on February 3 1942 six Centaurus Typhoon Is were ordered, *LA594, LA597, LA602, LA607, LA610* and *LA614*. Three days later Hawker told the MAP that they could modify a Typhoon airframe and convert Tornado *HG641* into prototypes, but requested the order be reduced to four aircraft, cutting their heavy work load. Sensing reluctance to place a Centaurus fighter at high priority, the MAP refused to agree. Six it must be, the first to fly in October and all to operational standard. Hawker then asked to build two pre-production aircraft, but the MAP also turned down that suggestion.

Hawker, busy with the Sabre Typhoon II, complained repeatedly about demands placed upon them. As a result the MAP cancelled the Centaurus Typhoon I on June 2, apart from two examples to give Centaurus experience. Nine days later the Typhoon II programme was completely revised. The six aircraft would now be completed as two each with Sabre IV, Griffon (Mk IIb until the 61 was ready) and Centaurus. Hawker at once stated that building the latter would be a long job. It could not enter production before late 1943 because they had been ordered to give the Sabre Typhoon priority. Hawker asked Bristol to help by converting another Tornado. They were too busy. Relations were not improved when, on July 29, the MAP told Hawker that the Centaurus Typhoon II now had top priority whilst the firm was desperately trying to speed the Sabre version. Camm retorted fiercely that Hawker were producing a Centaurus Typhoon II mock-up while being pressurised by others to build enough Typhoon II (Sabre) to equip two squadrons before March 1943. Hawker were increasingly concerned about their inability to do everything and at the ever-increasing weight of the Centaurus aircraft — well above that of the Sabre fighters which were no lightweights.

Lightweights? As the weight and difficulties in siting the Centaurus engine's exhaust system were bringing concern, details of F.6/42 reached Hawker. Camm's immediate thought was whether the Tempest II could somehow be adapted to meet it? His solution was to pull the wings into the body centre line, thereby eliminating the centre section interspar structure. He reckoned that improved pilot's view might be achieved by altering wing incidence, or re-shaping the nose and fitting contra-props.

On November 3 Camm and Rowe discussed this novel idea, and also the fitting of a Napier circular nose radiator combining reduced drag with less

Napier investigated several annular cooling systems, one on Tempest V EJ518 which joined the firm in May 1944, later that year served with 3 Squadron and suffered battle damage on January 6 1945 (Napier).

vulnerability to enemy fire. Balance, they agreed, might be a problem, an annular radiator adding 400 lb forward weight. If a contra-prop was also fitted, the engine would need shifting about three feet aft — the cockpit, too. Rowe suggested that Camm consider an internal radiator behind the pilot for the liquid-cooled engine variants after the style of the Mustang and a new Martin Baker design. Evidently that company had something to offer after all! Weight could be lost by using shorter-barrelled cannon, the Hispano Mk IV being a foot shorter, the Mk V even more.

Back in London Rowe commented that the Tempest II development 'looks promising'. A normal Tempest II would weigh 10,990 lb all-up, the development 10,300 lb. Rowe said that he 'found Camm quite receptive of ideas for improving the Tempest II when I saw him at Claremont on November 13'. He added, 'I told him that I thought he had approached F.6/42 ideally, reducing the fuselage to get a much better view, using contra-props and reducing the span and weight by pushing the wing into the fuselage'.

Forecasts suggested a 2,700 lb Centaurus would produce 1,360 hp at 25,000 ft. At a dry weight of 1,600 lb the Merlin 61 was giving 1,270 hp at 25,000 ft and raising doubts about the necessity of going for the Centaurus. These were allayed when, with + 8 lb boost, the test bed Tornado's engine (dry weight 2,690 lb) gave 2,300 hp at 5,500 ft and 1,940 hp at 16,750 ft, proving its low level capability. Prognostications surrounding Hawker's F.6/42 provided excellent ammunition for the MAP to use against the Air Staff, the Air Ministry and their Folland. They admitted that Hawker's design was larger and heavier, but maintained that it came from an experienced firm — and by November Hawker had devised a Sabre-engined version having wing radiators. The Air Staff view remained that the best F.6/42 would come from a new consortium which must be allowed to grow because the old firms were 'insufficiently enterprising'. They accepted Hawker's scheme as 'worthwhile', but 'irrespective of the F.6/42 fighter'. They wanted the Folland fighter.

December 9 1942 brought a fixed battle between the two sides. Captain Liptrot, MAP, pointed out that although the modified Tempest II was larger, its extra span had compensations. The CAS, saying they were not looking for a compromise, added that he did not think the impressive Fw 190 was such a design. Then he dealt with the idea of contra-props, which were not new. Indeed, Fairey's 24-cylinder Prince engine had them in 1939 although the gearing to produce two counter-rotating propellers driven by one engine, and devised to reduce swing, improve the holding of the line of flight (and aim) and keep down the ever-increasing size of propellers needed by powerful engines, was very complex to produce and maintain. Captain Liptrot stated that, once committed, the fighter could not change its habit. Then the Air Staff enquired about using a laminar flow wing. Liptrot replied that Camm was using a 'fully engineered' wing! As to formation of a new team, the MAP pointed out that it needed not only leadership but full, widespread support.

Summing up, the ACAST stated that the Air Staff 'accepted the Tempest development as a good thing, but still thought it was not good enough', and wanted a new type to back the improved Tempest. 'In the case of the Spitfire, modifications would keep it going for three to four years. New wing, new Merlins, Griffons; the Air Staff were glad to know all of this.' But no agreement on F.6/42 had been reached.

The Air Ministry took a tough line against the F.6/42 Tempest II development after the DTD's figures showed little difference between this and Folland's Fo 117. The Secretary of State for Air commented, 'I am in difficulty understanding the figures. It seems strange to me a design so unattractive that it had to be redrawn should possess the same speed and rate of climb as its rival. I agree, this modified Tempest is all right for early 1944 — glad we can have it in production then. I understand the Folland is a full 20 mph faster than shown in your letter. It is a new design; the Hawker is a re-hashed Tempest. Folland has shown ability to provide the best answer when a new specification was laid before their design staff. I want to keep them. I want the Folland design, constructed to F.6/42.'

'But their estimates are paper ones only', retorted Air Marshal F.J. Linnell, CRD. 'Seldom in the past have such estimates proved to be right. I'll be surprised if it weighs under 10,000 lb by the time it is built.'

Air Chief Marshal Sir Charles Portal, Chief of the Air Staff, then threw his

High hopes surrounded the Sabre engine fitted with an annular radiator. Typhoon R8694 *was thus fitted for trials* (Napier).

support behind Folland and, writing to Air Chief Marshal Sir Wilfrid Freeman, stated, 'I am very sorry to note from your last letter that you are still opposed to the construction of the Folland F.6/42. I am not in a position to dispute technicalities, but do wish to bring to your notice one point in favour of it. You say the weight advantage is not expected to be more than 100-200 lb. Maybe so, but the Folland has a contra-prop and the Tempest development does not. In view of improved handling qualities no doubt the Tempest will eventually have to be so fitted. Consequently, its weight compared with the Folland should be increased — by about 400 lb, I am told. Actual weight advantage of the Folland must therefore be about 500-600 lb.'

Folland's design had now been further updated, its top speed forecast as 474 mph at 20,000 ft against the original 458 mph at the same height. Taken into account was increased output from the Centaurus XII which might even have raised its speed to 488 mph. With the new engine the Folland started at an all-up weight of 9,170 lb when fitted with a six-bladed contra-prop and carrying 135 gal of fuel.

As heated retorts and counter-blasts raged elsewhere, Supermarine forged ahead with their new wing, claiming that it could also be fitted to the Spitfire IX. This was not strictly accurate, but on December 29 1942 the company was ordered to build three laminar flow wings for Spitfire VIIIs, whose speed was forecast as likely to rise by 8-10 mph, the first pair of mainplanes being due for delivery in eight months' time. An entirely new laminar wing Spitfire variant prescribed to Specification F.1/43 would need a 124-gal fuel load and provision for a 90-gal drop tank. Supermarine, well removed from the F.6/42 squabble, merely contented itself with mass-producing real aeroplanes, valuable Spitfire Vs and IXs, while planning Griffon successors.

Throughout the F.6/42 campaign Hawker was deeply involved with the Tempest V, projected 1943 production being for 90 aircraft. Camm accepted the Mk II (Centaurus) as the main version, yet the MAP persisted in prodding

Typhoon HG641 *fitted with a Bristol Centaurus IV, four-bladed propeller and a much modified cooling layout. Wing gun ports identify use of a Mk Ia airframe* (IWM MH4957).

Hawker to design various Sabre variants of F.6/42 and proposed both the 37 ft-span wing and that its area be reduced by ten per cent — to save weight, or face? Some indication of likely Tempest II performance came with detailed analysis of that of the Tornado *HG641* (Centaurus IV). This showed a top speed of 429 mph at 17,800 ft and 21,000 ft reached in 5 mins after taking off at 10,095 lb.

No Tempest II could enter service before well into 1944. That was clear by January 1943, due to the speed of development and production of the Centaurus and difficulties with installation of its exhaust system in a fighter. This raised the notion of cancellation in favour of the light version of the Tempest II which, to differentiate it from F.6/42, was now being referred to as OR 121 alias F.2/43 — and probably not to the pleasure of the Air Staff. Suggestions that the F.2/43 could fly in 11 months' time were viewed sceptically by the Air Staff so it was decided to continue the Tempest II and avoid, if possible, involving the F.2/43 with the Sabre.

On January 29 1943 the Tempest I prototype with wing leading edge radiators for its Sabre IV was transported to Langley where its first flight took place on February 24. Available performance figures for its companion, the Tempest V (flying weight 10,960 lb), showed a top speed of 420 mph at 19,000 ft. It took 2.75 mins to reach 10,000 ft and 6.35 mins to 20,000 ft.

Tempest progress was reviewed at a CRD meeting on February 3. Yet again argument ensued over the Folland's merit. Liptrot tried a new play — Hawker estimates were based upon flight trials whereas Folland's were but paper estimates. He added, 'There has been high pressure salesmanship since January 5, with Folland's claiming 476 mph at 20,000 ft — 20 mph faster than the revised Tempest. I told Mr Ratcliffe [of Folland] quite bluntly that vague, impressive performance claims cut no ice with us whatsoever, that he would have to present material to justify them.'

Pressures to halt Folland's brave incursion into the fighter field had become enormous. Exactly how it was ended one may but surmise. On March 17 1943 Air Marshal Linnell wrote to the ACAST, 'As you know, meetings whether to build or not have been going on a long while. I have received instructions today from Controller Engines that Folland's should be told not to build such a prototype.' Portal was furious and told him to refer to recent correspondence. 'On January 10 1943', he said, 'I placed on record my serious protest against any intention to abandon the Folland.' A further letter, of March 24, stated, 'Sorley tells me it was agreed at a meeting on March 18 that, compared with the Hawker, the Folland design gives a better rate of climb of 400 fpm, and higher maximum speed by 10 mph assuming the same propeller efficiencies. I am sure we cannot afford to forgo these advantages. I feel more than ever we must introduce the Folland fighter.'

Next day the CRD replied, 'I am still unconvinced about the wisdom of Folland doing it, they are too weak. The Admiralty 28/40 design would have to be taken away from them. Of the other two projects — 6/42, 2/43 — we'll have the latter months, if not years, before the Folland. Where would its production take place? The aircraft cannot effect the war effort before the end of 1944.' And so the Folland fighter was killed although English Electric tried to revive it to F.19/43..

Meanwhile, the Griffon Spitfire XII was ready for service. Fortunately, it did not suffer the buffeting which clipped wings induced on Spitfire Vs. AFDU was ordered to compare the clipped and non-clipped wing Mk V and found the LF V

Great hopes were pinned on the Sabre IV Tempest I whose radiators were finally placed in the wing roots (IWM MH4950).

to be 5 mph faster at 10,000 ft, no speed difference between 15,000 and 20,000 ft and the normal aircraft slightly faster at 25,000 ft. Acceleration was faster with the LF V whereas, between 20,000 and 25,000 ft, the unclipped aircraft climbed at 15 secs-a-min faster. Diving, the LF V drew away, its aileron response quick and crisp. It could turn on a normal Mk V in 20 secs using superior rate of roll. Perhaps most important of all, its turning circle was 1,025 ft at 20,000 ft compared with 1,450 ft taken by the Fw 190, although clipped wing tips upset the airflow over the tailplane.

HM599, the Tempest I prototype, made two flights on February 24 at a weight of 11,050 lb. Handling was superior to the Mk V's although loss of elevator control was noted. Prior to the flights the Sabre IV's need for a larger radiator was certain. Fitting sliding canopies to Tempests was expected to reduce vibration. By March 1943 1,138 Sabres had been built and 326 were unserviceable. By May, 900 of the 1,338 constructed were being repaired or modified at Acton or Aldenham. Even more disturbing, Napier could cope only with 70 a month whereas Gloster were building 33 Typhoons monthly. Little wonder there were suggestions that Gloster switch to producing Spitfires and provision of extra factory space for Sabres was halted. None of this boded well for the Tempest.

Early 1943 found Supermarine busy with more Spitfire improvements. Middle East Command had already requested a low-level Spitfire to combat Fw 190 fighter-bombers. Cropped impellers and clipped wings were recommended, increasing the speed at 2,000 ft by 20 mph. The Air Staff wanted the MAP to ensure that all 1943-built Spitfires would have only cannon, but this was never achieved. Although able to accommodate four cannon, Spitfire Vcs usually carried two (except when overseas) due to inadequate gun heating facilities.

Supermarine's design and development interests were increasingly switching to Griffon Spitfires. That engine had undeveloped potential whereas the Merlin might soon be worked out. Model tests of the F.1/43's wing were now under way, the RAE and the NPL both trying to discover the likely landing run of this laminar flow wing Spitfire. Careful aileron design was very important, also radiator siting and cooling flaps. Clearly, dihedral would be important too and was provisionally set at 3°. The pilot's improved sight line over the nose was to be at least 5°. Supermarine forecast the first flight of the wing on a Spitfire VIII as likely to be July-August 1943, but already Specification F.1/43 covering the new design was being amended to call for a much altered aeroplane, not merely a modified Spitfire. Apart from nose and wing alterations, it was to have a wide track, inward-retracting undercarriage. Also under development was a new line of two-stage Griffon-engined Spitfires based upon the Mk XX derived from the Mk IV. The single-stage Griffon-engined Spitfire XX fell by the wayside because the Mk IX achieved as much as the Mk XX. First would come the Spitfire XXI (Griffon 61) which virtually slipped into being in mid-March 1943 with no formal contract calling for a Spitfire to this new standard produced by wedding a Griffon 61 to a Mk VIII airframe. A provisional order for 1,207 such Spitfires was placed on March 4 1943 and soon extended to embrace another 300 from the Phillips & Powis works at South Marston which Vickers was soon to acquire. Reduction of the huge total came on March 30, when only 562 Spitfires were required 'based upon the Mk XX'. By that time it had been agreed that an interim F.1/43 would be produced by mating the laminar flow wing directly to the Mk VIII's fuselage and fitting a 24-volt electrical system.

Throughout the winter, testing of Merlin-engined Mustangs was continued, results being discussed at Hucknall on February 2 1943. AFDU reported that *AM208* proved unacceptable in its present form. Insufficient directional stability was recorded, and there was a large change in directional trim with power — objectionable due to heavy rudder forces involved. While manoeuvring, the aircraft easily slipped sideways. Propeller size played a part, but A&AEE representatives suggested a booster tab be fitted to the tail spring tab. A fin fillet fitted on to *AL975* certainly helped to cure instability. Boscombe Down's trials of *AM208* (Merlin 65) at a take-off weight of 8,650 lb showed climb to 14,000 ft in 4.2 mins at combat power and 5.5 mins using normal power. Maximum speed was 406 mph at 12,000 ft MS and 433 mph at 22,000 ft FS. The engine used failed later and speeds recorded with its replacement were lower: 395.5 mph at 9,800 ft MS and 425 mph at 21,400 ft FS. Estimated service ceiling was 37,700 ft. The Merlin Mustang was not as outstanding as had been hoped.

Change had already come to its production programme. Under Lend-Lease arrangements 550 complete Merlin Mustangs were to be delivered from the USA before another 1,000 came to Britain to receive Merlin 66 engines. This rendered the five development prototypes unrepresentative of production machines. The new production aircraft were, however, designed for the Merlin 68 and had a fuselage 2 in deeper, giving a clearer run to internal piping. North American had run into severe corrosion problems with the Merlin 68 engines and Rolls-Royce suggested that it was due to high radiator resistance. A pressurised system was the answer and Rolls-Royce carried out the remedial research for Packard.

April brought results of Tempest I tests, its top speed being 458 mph at 25,600 ft. At 29,000 ft it reached 451 mph but was not carrying any operational load. Such was not the case with the Spitfire XII, the first operational Griffon version which entered squadron service on February 24 1943 and became operational on April 16.

As the flow of Spitfire IXs increased it was suggested that all Mk Vs have wing tips clipped and impellers cropped. Eventually only some of them were modified, the remainder being held for medium level operations and bomber escort. They fulfilled a most useful role when the USAAF brought B-26 Marauders to Britain and needed them escorted.

Somewhat unexpectedly, and another interim variant, came the excellent Spitfire XIV. As Merlin 45s upgraded the early airframe and the Merlin 61 another, so the trend continued. Now a Griffon was fitted to a Mk VIII's airframe, bridging the gap prior to laminar flow wing Spitfires. *JF316,* first flown in January 1943, was the first converted, its armament restricted by weight limits to two 20 mm cannon and four .303 machine-guns. The conversion achieved 445 mph at 25,000 ft and a decision to put the aircraft into production later in 1943 as the Mk XIV was taken on April 29. Six Mk VIIIs were converted for Griffon 61 trials and an initial order for 50 Mk XIV's powered by no-frills Griffon 65s — similar to the Mk 61 but with reduction gear ratio changed from .451:1 to .510:1 — was placed on May 25, 1943. Engine power, driving a strange-looking five-bladed propeller, could accept a boost of + 18 lb/sq in. A 45-gal drop tank would be standard, and all aircraft would be tropicalised. The aim was to have the Mk XIV operational in the shortest possible time, and it was hoped that later aircraft would have the 'teardrop' rear view canopy.

Construction of the Tempest II was very slow, in contrast to the success of the Centaurus Tornado. Of its 100 trouble-free engine hours, some 50 had been logged in one week. Folland's (23/37) Centaurus engine test bed was about to fly and Bristol hoped to have an engine available for the Tempest II late May 1943. The second Tempest II was far behind the first, partly because of Hawker's reluctance to fly it before trials with the first were well advanced. To speed development the fourth production Mk V was to be completed as a Mk II. Further reluctance over the Mk II lay within its relevance to the Hawker P.1026 F.2/43, the specification issued on May 14 1943 for a light fighter now named 'Fury'. Powered by a Centaurus XIISM driving a five-bladed propeller, the aircraft would be a mixed load fighter-bomber, although its operational height was listed as 30,000 ft, not the expected 20,000 ft. An 8° view forward was specified and a spring-mounted seat to reduce fatigue. Four Hispano Mk Vs would be fitted, the airframe being able to safely dive at 525 mph EAS. For faster delivery

JF318, a Spitfire XIV prototype with interim type of fin and rudder.

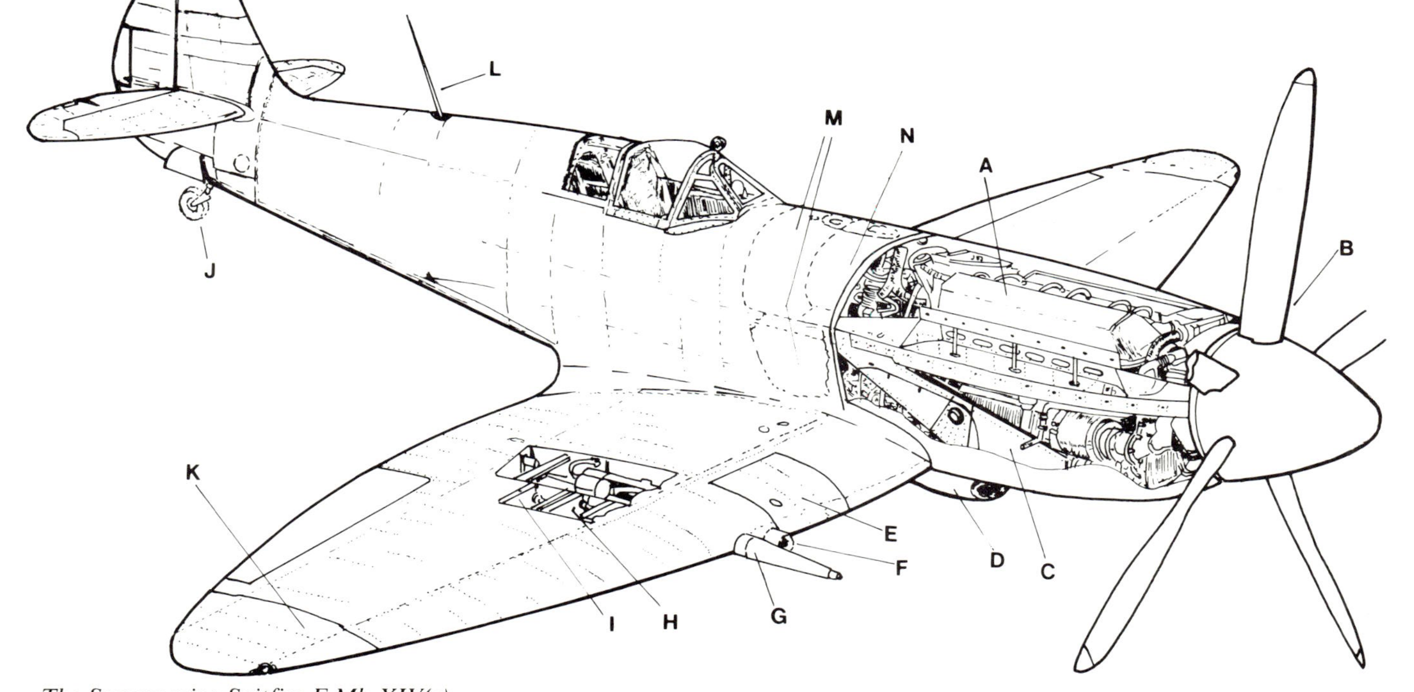

The Supermarine Spitfire F Mk XIV(e).
Key: *A — Rolls-Royce Griffon 65 (2,035 bhp low gear, 1,820 bhp high gear both at 2,750 revs + 18 lb boost, former at 7,000 ft, latter at 21,000 ft. 12 cylinder 60° Vee, dry weight 2,090 lb), B — five-bladed Rotol Jablo R4.19/5F5/1 propeller of 10 ft 5 in diameter and 35° pitch range, C — Engine mounting beam, much strengthened for the Griffon, D — air intake, E — fuel tank (in each wing) of 12¾ gal, F — .5 in Browning gun sited in position feasible for a second cannon, 250 rpg, G — 20mm Hispano Mk II, 120 rpg, H — ammunition store for machine-gun, I — ammunition store for cannon,*

K — detachable wing tip to allow for bolting on of wooden fairing for clipped wing variant, J — retractable tailwheel, L — whip aerial, M — top fuel tank (36 gal) and lower tank (48 gal), N — oil tank.
Dimensions: *Wing: span 36 ft 10 in, gross area 242 sq ft, tip height above ground (tail up) 5 ft 8½ in, section NACA 2200 series, incidence 2° at root, dihedral 6°. Fuselage length 32 ft 8 in. Height (one blade vertical) tail up, 11 ft 1½ in, tail down, 12 ft 8¼ in. Undercarriage track 5 ft 8½ in. Tailplane span 10 ft 6 in, width 4 ft 1 in. Wingspan, when clipped, 32 ft 2 in.*

the first of two prototypes ordered in June 1943 would have Tempest outer wings and ailerons, the second the redesigned ones.

While the first production Tempest Vs were being completed in May the Sabre position remained alarming. Napier had prepared the Sabre V for the Tempest I, a complete revision which showed some promise. Development took place at Acton, but the Sabre V did not find favour until 1945. Attention reverted to the Mk II with attempts to cure defective sleeves and the slow speed clutch disc on the starter which plagued many engines. Napier accepted Group Captain Banks' suggestion that the Acton factory should concentrate upon repairs and modifications and not more development and production. By June 1943 the far from healthy Sabre IIa had appeared. Whereas the Mk I, giving 2,060 bhp at 3,700 revs at 2,750 ft and 1,870 bhp at 15,000 ft, could accept only +7 lb boost, the Mk IIa was able to take +9 lb boost for 5 mins at combat rating. The later Sabre IIb was suitable for +11 lb boost and either three- or four-bladed propellers. Further development resulted in the Mk IIc, the previous low gear ratio of .448:1 being increased to .473:1 in this mark. High gear ratio was set at .626:1 in all Sabres so far, including the Mks IIc and IId both of which took four-bladed propellers and gave their best low altitude rating at 6,500 ft instead of 4,750 ft. Net dry weight of the Mk II series was 2,380 lb, an increase of 150 lb over the Mk I.

Of the Sabres in general Group Captain Banks commented, 'No amount of engineering skill could overcome results of sheer carelessness and bad work of operatives. Some failures are attributable not to legitimate wear and tear but to the fact that foreign matter, including machine trimmings and broken piston rings, has been left in the body of the engine during assembly.' No aircraft were known to have been lost due to this. 'On balance', he wrote, 'the engine is a miserable failure.'

Despite its Sabre, the first production Tempest V, *JN729,* flew on June 21 1943. This, and other early examples, had long-barrelled Mk II cannon and were designated Mk V srs i. Carrying 132 gal of fuel, similar aircraft *JN731* weighed 9,057 lb tare and 11,484 lb operationally loaded. Its top climb rate fully loaded was 3,815 fpm at 3,500 ft in MS gear; 2,680 fpm at 15,800 ft in FS gear. *JN731* took 2.9 mins to reach 10,000 ft; 6.85 mins to 20,000 ft and 14.3 mins to 30,000 ft; service ceiling being 34,800 ft. Applying +9 lb boost raised the maximum speed to 411 mph at 6,600 ft (MS) and 432 mph at 18,400 ft (FS). Cruising at 220 mph IAS at 10,000 ft allowed a range of 675 miles and a duration of 2.8 hrs. Early flight trials showed a need for lighter ailerons and improved elevators.

Powered by a rigidly mounted Centaurus IV, the prototype Tempest II *LA602* (all-up weight 11,350 lb forecast speed 440 mph at 22,000 ft) first flew on June 28 1943. Ideally it needed a Centaurus XII, but N.E. Rowe told Hawker that only five such production engines could be ready by September 1944, production for Tempests being impossible prior to 1945. Tempest I *HM599* proceeded to A&AEE late August 1943 for full load speed trials, the best recorded being 439 mph against 455 mph forecast — in addition to which the aircraft was a year late. Its speed was about to be bettered by Spitfires. Supermarine was building the F.1/43's wing for what would be basically a Mk VIII's fuselage. A first flight remained some way off for the new wing which was expected to be able to cope better with compressibility effects and satisfy 'a very insistent demand from the Air Staff for a considerable improvement in the

The prototype Tempest II LA602 photographed during summer 1944. Originally it had a Typhoon-type fin. Its exhaust system differs from the production type.

roll ability of the Spitfire. Any performance gained in other directions would be useful but incidental', Supermarine decided. Manoeuvrability emphasised in the F.1/43 Specification included 2.03 secs to bank 45° at 400 mph EAS at 10,000 ft. Supermarine maintained that the design was based on an ability to make a 90° bank in one second as the MAP made it clear that the aircraft would be unacceptable if it failed to meet the requirements.

Merlin Spitfire development was virtually ended now. No Mk Vs were built having clipped wings, but the parts to permit that modification were mass produced. June 1943's decision that 75 per cent of production Mk VIIIs and IXs must have Merlin 66 engines would give between 15,000 and 25,000 ft performance. So strident had the call for Griffon Spitfires become that in August 1943 action was taken to allow additional Mk VIII airframes to take Griffon 65s.

Specification 21/P1 covering two-stage, Griffon 61 Spitfires was released on August 2 1943. Production would commence at Castle Bromwich in January 1944 and because of many modifications, under a new nomenclature — Victor F Mk I. The laminar flow F.1/43 would be the Victor Mk II. Service entry of the laminar flow Mk VIII was scheduled for August 1944, the Victor I for November.

Already the programme had been upset when the prototype Griffon 61 Spitfire, *DP851,* first flown in December 1942, was involved in an accident in May 1943 after Supermarine had recorded a top speed of 455 mph at 25,600 ft and climb to 30,000 ft in 7.51 mins. Its place was taken by *PP139,* the Mk 21 prototype, first flown on July 24 1943. Whereas *DP851* had limited structural strengthening this new aircraft had numerous revised features. Like the intended production Mk 21, it featured a much stiffer wing able to accept stronger ailerons permitted by their attachment through 'piano hinges'. More fuel would be carried, also a permanent load of four cannon and protective armour. A stronger, lengthened undercarriage allowed for the 11 ft 7 in diameter five-bladed propeller.

Another important stage in Spitfire development at this time was the 'tear-drop' canopy and cut-away top rear fuselage first in evidence on Mk VIII *JF299.* On August 8 1943 it was delivered to Wittering where AFDU commented

Spitfire VIII JF299 *was the first Spitfire to have a rear view canopy fitted* (via Bruce Robertson).

most favourably upon a feature that was common in 1945 but which could not have been introduced much sooner without considerable disturbance to production lines.

High output from the Griffon and increased take-off swing brought difficulties to the latest Supermarine products in the summer of 1943. One answer was a new propeller which would also be of benefit to the F.1/43. While four-bladed propellers remained satisfactory, five blades were desirable.

To counteract forces being produced, fin and rudder areas were increased, this requiring experimentation and skill. For better control of an aircraft far weightier and more powerful than originally conceived, counter-rotating propellers were attractive because of the positive effects of their increased blade surface area. Following contra-prop tests in July using a Spitfire IX, Rolls-Royce reckoned Spitfires needed an additional 4 sq ft of fin area. A contra-prop Tornado had additional fin area of 1.8 sq ft, then 2.2. Considerable weight increase of contra-props brought problems for fighters, although handling advantages were much improved. As for speed, a Spitfire LF VIII with all-up weight of 8,050 lb (compared with 7,900 lb with four-bladed propeller) and powered by a Merlin 70 driving a six-bladed contra-prop, reached 419 mph — only 2 mph faster than normal. Contra-props raised the Mk XIV's speed by only 4 mph so the five-bladed propeller with blades slightly widened was retained, offering increased solidity. As the latter became greater, so did torque reduction. There was a better 'feel' to contra-prop aircraft in normal flight, their propellers aiding rapid trim if an engine cut, and when the throttle was opened slowly. These reasons prompted A&AEE to advocate, strongly, contra-props for the F.1/43. No contra-prop trials had been undertaken there by October, but the Tornado had flown 32 hrs by early October and a Rotol contra-prop Spitfire (Merlin 61) 21 hrs.

Results were available of trials with several contra-prop Spitfires. Mark XIV *JF321* (Griffon 85) had varying rudder horn balance areas tested in attempts to increase rudder effectiveness and tighten turns. *AB505* (Merlin 63), a Mk IX, showed excellent control and good stability — much better than the five-bladed propeller on the Mk XIV induced. *MA587,* a Mk IX, was set aside for contra-prop endurance trials, and *MH874* was being modified for contra-props at Castle Bromwich to have a Merlin 63A or 64 as the propeller's gears could only fit an aircraft with a Coffman crank case.

For the present the RAF would have to accept the Spitfire XIV with its five-bladed propeller, service release being approved on September 17 1943. The RAF was thus getting a second interim Griffon (65/RG4SM) Spitfire carrying 85 gal of fuel in the fuselage and 25 gal in wing tanks. Its fin had a 4 sq ft area increase over that of the Mk V, and the rudder, on Mk XII experience, was further enlarged. *RB140,* the first production aircraft, needed clearing at Boscombe Down in October. Its weight limitation was 8,500 lb, maximum diving speed 470 mph and armament limit two 20 mm Hispano Mk IIs. Prototype *JF319* was raised to production standard, with enlarged radiators, to speed these trials, for plans had called for six production Mk XIVs to be built by November. First deliveries were to 610 Squadron in January 1944.

All work with very powerfully engined Spitfires indicated needs for careful tail unit design and in mid-September 1943 Supermarine proposed a still larger fin and rudder for the F.1/43 because its cg was further forward than the Spitfire's. Planning was for it to have a Merlin driving an 11 ft 8 in propeller with demand for contra-props ever increasing. The problem was not only of their production but, also one of suitable engines. Rolls-Royce were scheduled to produce 438 non-contra-prop Merlins between November 1944 and August 1945 and on November 18 1943 categorically stated that they had no capacity to build any other version. The MAP weighed in with a reminder that current policy was to equip only two Spitfire 21 squadrons with contra-prop aircraft, their engines to become available in 1944. Better, they felt, to tread with caution in view of the added complication of maintaining contra-prop aircraft.

F.1/43 delays were increasingly disturbing. In November 1943 Supermarine shocked the MAP with the revelation that only if more labour could be found might a prototype emerge in March 1944. This, they proposed, should have a Spitfire XIV's fuselage, Griffon 65 and a contra-prop — if available. The second and third prototypes would have modified Mk XIV fuselages incorporating forward view requirements, improved rear view and 24-volt electrics. A Griffon 61 with contra-prop would be fitted to the second aircraft, equating the

Interim versions of the Spitfire proved highly successful, none more so than the Mk XIV, of which RB142 was the third production example (RAF Museum P9148).

Sharing Hucknall with Merlin Mustangs, Spitfire IX JK535 used for contra-prop trials (Rolls-Royce).

standard of the first production machine due in August 1944. Contra-props would be fitted to the third, Merlin RM14SM-powered.

Development of the laminar flow wing had taken so long that the Air Staff and the MAP now decided that it would not be introduced on Mk VIII and Mk XIV Spitfires. On November 19 the MAP wrote to Supermarine expressing deep concern that the prototype was so delayed, reminding Joseph Smith that he said this 'was not due to lack of design staff', adding that his experimental team did not agree. Supermarine representatives went to the Ministry on December 10 1943 to explain the situation, Smith attributing delay to a late request that the fuselage be able to accept a Merlin or Griffon yet retain better forward view, to which had recently been added a request for a cutaway rear fuselage. He did, however, admit that the time to be taken for the task, had, in August 1943, been underestimated and suggested March 15 1944 as a likely first flight date. On this assurance, the MAP ordered a fourth, navalised prototype in an attempt to encourage Supermarine to work faster.

Amidst confusion as to whether both F.1/43 variants would proceed, with or without contra-props, it was decided that a meeting — to be held on December 29 1943 — should discuss all the Supermarine fighters. The decisions taken were that two Mk VIIIs would have contra-props and two Spitfire 21s would have them for 150-hr type testing. In March 1944 a Mk VIII would have a de Havilland contra-prop fitted and between March and June a dozen F 21s would get them. Contra-prop No 19 would, in June 1944, become available for the second F.1/43. All this assumed that sufficient gears were built, but by January 7 1944 it became clear that production of gears for Merlins could not possibly be undertaken that year, which made the Merlin 1/43's future bleak.

Hawker, in autumn 1943, were busy with the light Tempest, the F.2/43. Estimates for a two-stage Griffon RG15-powered version reached the MAP on September 20, sanction to proceed being given on October 29. Projected figures included a service ceiling of 41,000 ft, maximum speed of 470 mph at 12,000 ft, climb to 20,000 ft in 5.7 mins and a 595-mile range able to be extended to 975 miles with drop tanks. This compared with the forecast Tempest II's maximum of 440 mph at 20,000 ft and service ceiling of 37,500 ft. Reduced wing span of the 2/43, from 43 to 38 ft and lowered wing area from 306 to 205 sq ft, would cut the all-up weight by 1,000 lb. A Centaurus XIISM instead of a Mk IV would raise additional power around 7,000 ft and a forward downward view of 8° was likely. To the front bulkhead the fuselage would be monocoque, instead of tubular between the pilot and engine as featured by the Tempest Mk II. Hopefully, this would reduce vibration met with in the Tempest. Its overbalance and snatch problems might also be avoided and the rate of roll be improved.

Demands for the Tempest II in the Far East were ever-increasing. Hawker were eager to strengthen its fuselage to avoid any repetition of the Typhoon's problems, but the effect upon range of every additional item worried the officials. Currently the aircraft had a 250-mile operational radius. More fuel additional to the 160 gal permanent tankage might be afforded by using a 90-gal belly tank, freeing wing racks for weapons. Its wing guns had 175 rpg compared with the F.1/43's and Spitfire 21's 150 rpg. Before the Tempest II could be accepted for service, however, excessive vibration, which the engine mounting passed on to the airframe, had to be cured. Carburation problems above 20,000 ft added further delay, also a need for better oil circulation. In the case of the F.2/43, a rigidly mounted Centaurus XII was all that was available pending the Centaurus XXII, expected in October 1944, and planned for a flexible mount. Since a Griffon would be ready by August 1944 the plan of December 1943 was revised to call for the second F.2/43 to have that engine, the third machine to be a Centaurus XII navilised example like the fourth. Possible use of a Rolls-Royce H46 was discussed but rejected.

Tempest II service trials were scheduled for March 1944, but it was far from ready. Hawker feared an unpleasant report upon their forthcoming main production type. *LA602* had 'Siamese-twin' exhaust pipes, modifications to which were reckoned inadvisable. Only a mock-up of the single stack exhaust system had yet arrived. *LA607,* the second prototype, first flown on September

The second prototype Tempest II LA607 survives in the USA, after being preserved in several museums in Britain. Here it is on display at Cranfield.

18 1943 and grounded after engine failure, was being rebuilt with single outlet solid drawn pipes in a flexibly mounted engine. Clearly, Tempest II production would increasingly become retarded unless rubber-mounted power plants became available. March brought the disturbing news that the flexible engine mounting could not accept the Centaurus XII without extensive redesign — and there was insufficient time for that. Similarly affected would be the F.2/43. On March 20 1944 the MAP informed Hawker that Tempest IIs would, unfortunately, have the Centaurus V.

Two squadrons were already flying Tempest Vs, the first being No 486 to which delivery commenced on January 14 1944. Late in February No 3 Squadron began to equip, but not until late June was a third squadron, No 56, flying Tempest Vs.

As the major contractors engineered their latest designs, Martin-Baker, whose creative ability passed largely unrecognised, were busy. Following the loss of the MB 3, that design was re-engineered as the Griffon MB 4 and unsuccessfully tendered to F.6/42. Official pronouncement that the MB 3 was outdated before it flew led an update of the MB 4 in 1943 to include a four-cannon wing in line with revised Air Staff policy. A teardrop canopy and rear fuselage radiator gave the aircraft a similar appearance to the Mustang. Fuselage fuel tanks held 200 gal which gave the aircraft a range of about 1,240 miles at about 230 mph. The MB 5 was first flown from Harwell on May 23 1944, flight trials showing a top speed of 460 mph at 20,000 ft. A 2,340 hp Griffon 83 driving a six-bladed Rotol contra-prop conferred upon the MB5 a superb performance matching excellent handling qualities which were to be shown a number of times in postwar flying displays. Its development by a small concern working on a shoestring basis was indeed amazing. Such activity was not within the usual guidelines for fighter development and the firm was very badly treated. After learning in December 1943 that cg problems had caused the engine to be positioned further forward, and for that reason a contra-prop had become desirable, the CRD commented 'Martin-Baker keeps fiddling with the aircraft, contracts being meaningless. All I can do is push him along and hope we get it into the air one of these days. I don't think there's any loss.' In that

Production Tempest Vs had a fin of increased area, compensating for the enlarged nose radiator (H.A. Kofoed).

Mustang IIIs served in the RAF mainly as long-range escort fighters, but FX893 *shown here displays the type's ground attack potential* (RAF Museum P12052).

respect he was wrong, but Martin-Baker had taken too long to produce the MB 5 because they lacked the facilities — and sufficient government support. Unlike North American.

First shipments of their production Merlin Mustang IIIs had arrived in Britain in late 1943. Trials were eagerly undertaken to assess just how good the combination had become. *FX953* with a Packard V-1650-3 Merlin and a take-off weight of 9,200 lb showed a top speed (FS) of 450 mph at 28,000 ft and, using MS gear, 424 mph at 15,500 ft. Its service ceiling was 40,700 ft. To reach 20,000 ft took 6.10 mins and to 40,000 ft it took 26.25 mins. Four .50 in guns (800 rpg) were wing-mounted. Better results were recorded with *FX858* (Rolls-Royce Merlin 100/RM14SM), the maximum speed (FS) being 455 mph at 17,800

A bright performer, the Martin Baker MB 5 which performed extremely well but arrived too late (Martin Baker).

ft, climb to 16,000 ft taking 3.2 mins, to 20,000 ft 6 mins and 38,000 ft 12.9 mins. During December No 65 Squadron became the first to equip with Mustang IIIs, followed by No 19 in January 1944, the two squadrons on February 15 1944 operating them for the first time during a sweep over Belgium. To the end of the war the RAF usually employed its Merlin Mustangs as long-range sweep and escort fighters, and by May 1945 had 11 home-based Mustang squadrons. Delivery of the Mustang III (P-51B/C) totalled 930, and the RAF commenced receiving Mk IVs with teardrop canopies in September 1944; a total of 232 coming on charge. Loaded weight of a Mk IV, with additional 85-gal fuselage tankage, was 9,855 lb. As with British fighters, attempts were made to produce lightened variants and one tested in Britain, *FR409*, an XP-51F, had an enlarged canopy although it suffered from large changes in directional trim. In interceptor state its loaded weight was 7,855 lb.

Production of Spitfire 21s commenced at Castle Bromwich in January 1944 but a year was to pass before the aircraft was accepted for squadron service. *LA187* (Griffon 61), the first production Mk 21, first flew on March 15 1944. Its features included a 'Mk XIV type tail', and pointed wing tips. Handling proved far from good and *LA187*, although fast, was slower than the prototype. On the other hand the Mk XIV was proving to be excellent. *RB140*, the first production machine, was delivered on December 20 1943 and 610 Squadron commenced equipping on January 2 1944, by which time ten Mk XIVs had been delivered. Response to the XIV's success came on February 2 when an order for a further 400 was placed. These, strengthened examples, would be able to carry 1,000 lb bombs or three 300 lb rockets under each wing. They were to have Griffon 69s,

The increase of Spitfire speed — with the Meteor III's for comparison.

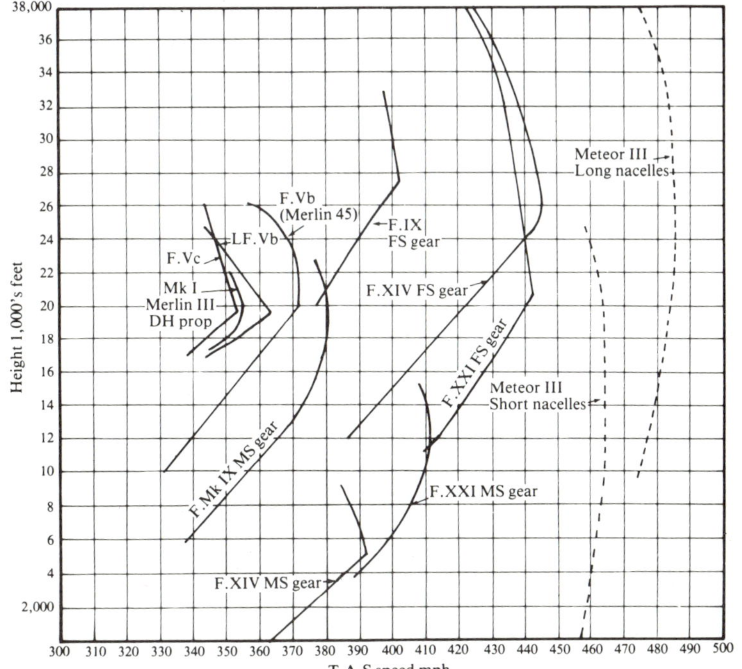

short-barrelled Hispano Mk Vs and maximum fuel capacity of 268 gal including use of drop tanks. Supermarine were informed on January 4 1944 that the DTD considered that five-bladed propellers were satisfactory for all 1944-built Griffon Spitfires, clearing away the contra-prop aspect.

More official discussion of the F.1/43 came on February 21 1944, its production scheduled to begin in October-November 1944. Laminar flow wing aircraft, it was suggested, should all be Castle Bromwich-built. Merlin 66s, now rejected for any F.1/43, would be fitted to a further batch of Mk IX Spitfires. Installing the Merlin 66 in the F.1/43 meant altering the under mounting to clear the carburettor. It was easier to fit all F.1/43s with Griffons and for Castle Bromwich to build only Merlin 66 Spitfires. Griffon Spitfire 21s could replace them fully in August 1944, and later have contra-props, leaving F.1/43s having five-bladed propellers.

Vickers explained plans for a prototype Spitfire VIII fuselage with a Griffon and contra prop and two F.1/43s (Type 371), one with Griffon and contra-prop and another with a Merlin RM14SM and contra-prop. The latter being in short supply, Supermarine were told they could only have one for F.1/43 trials. Already the Spitfire 21 was reaching far higher Mach numbers than thought possible and entering zones of considerable air compressibility. It needed a new tail unit before its limiting speed could rise from 500 to 525 mph EAS.

Further detailed considerations of the F.1/43 came on March 8 1944. The Type 371 wing would be based upon that of the second prototype which equated F.1/43/P1/SU (OR120). Powerful engines, high propeller tip speeds and compressibility effects causing shock waves to move across wing and tail surfaces led to wind tunnel investigations at the RAE and the NPL throughout 1943. Whilst laminar flow wings retarded drag increase, it was clear that very smooth skinning and surfaces were equally essential. Production F.1/43s were to accommodate 133 gal of fuel internally, carry 30 or 45 gal drop tanks, mount four 20 mm Hispano V guns and still be powered by the Griffon 65 or Merlin 14SM/16SM, since the contract had not been amended. Take-off must not exceed 1,400 yds. Supermarine claiming that the best performance came with a Griffon 65, it was decided all laminar flow wing aircraft should have that engine, policy to be reconsidered if three-stage Merlins became available before 1946. All advanced Merlins were already earmarked for Windsor bombers.

Within days of the decision there was much concern at Supermarine when it became obvious that Merlin F.1/43s would have much better range, essential in the Far East. The MAP, knowing none of this, reiterated on March 21 that all F.1/43s would have Griffons and that a few might later be converted for Merlin trials. Supermarine then, in a turn about, claimed that the Merlin version reaching 428 mph at 21,000 ft would be faster because it was lighter than the Griffon version with a forecast top speed of 417 mph at 26,000 ft. The CRD would not agree to any change. What disturbed officaldom more was the slow progress with the aircraft which, it was thought, was due to Supermarine's small experimental section. Air Marshal R.S. Sorley wanted to encourage the firm with an order for ten F.1/43s. Supermarine claimed a bottleneck with supply of hydraulic jacks, but the MAP thought there were other factors. Supermarine advised the Ministry that the prototype should fly mid-May and the second example in July 1944. Against this news the F.1/43 was named Spiteful on May 5, and the contract still listed the RG4SM and Merlin 16SM as suitable power plants. Two 250 lb bombs must be carried under the wings and permanent fuel

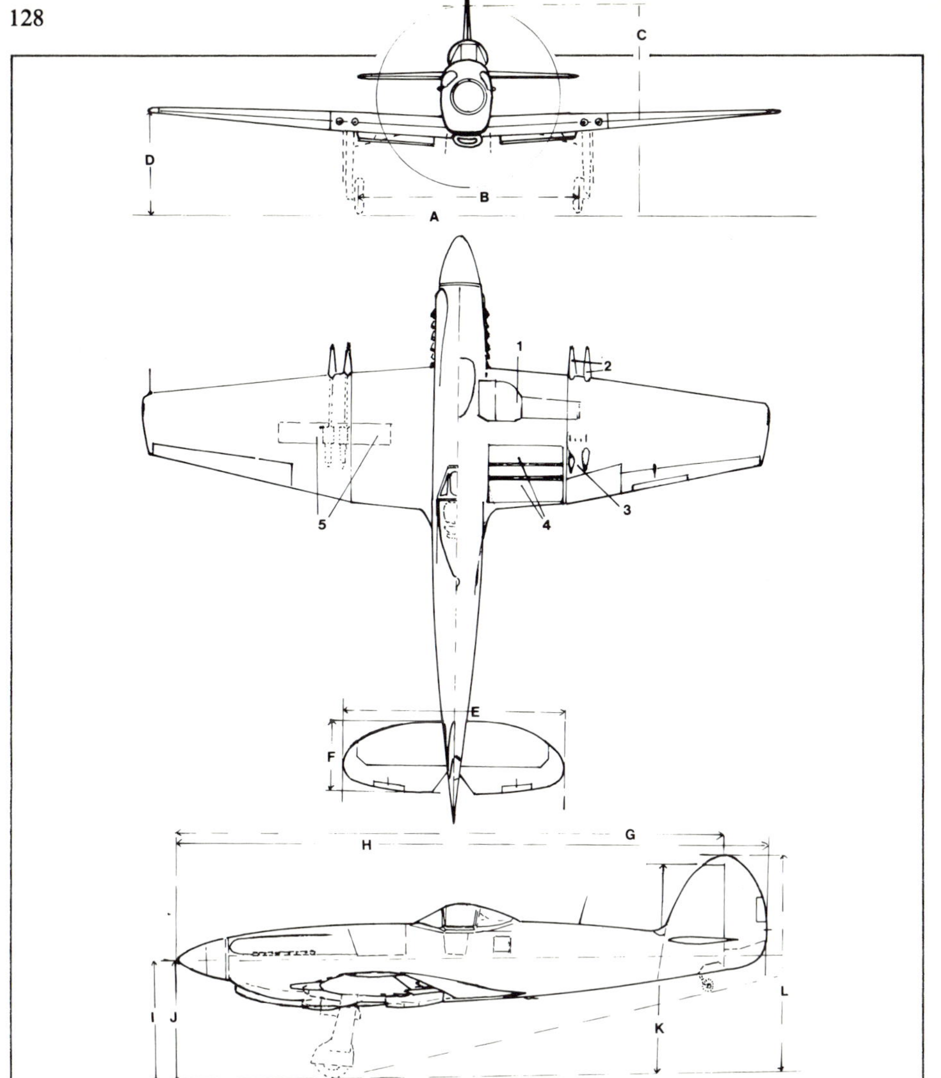

General arrangement of the Spiteful XIV.
Key: *A — wing span 35 ft 0 in, B — 12 ft 5½ in, C — tail down, 13 ft 5 in; up, 11 ft 11 in, D — 5 ft 4⅜ in, E — 12 ft 10 in, F — 4 ft 1 in, G — 30 ft 5½ in, H — 32 ft 11¼ in, I — tail down, 7 ft 7½ in, J — tail up, 6 ft 5 in, K — 12 ft 5 in, L — 12 ft 11⅜ in. 1 — inward-retracting undercarriage with closing door, 2 — Hispano 20 mm Mk V cannon; outer guns 145 rounds each, inner 116 rpg, 3 — fairings over gun breeches, 4 — radiators beneath each mainplane, 5 — panels over ammunition stores.* Further details: *Wing: aerofoil = Supermarine, High Speed 371/I and II, incidence 1½° ± 15', gross wing area 210 sq ft, dihedral 5° ± 15'. Propeller Rotol R1.4/5F5/F2, five-bladed, 11 ft diameter, pitch range 35°, constant speed. Top fuel tank 34½ gal, bottom 43 gal, two saddle tanks each 11¼ gal, two interspar wing tanks 8 gal each. Later aircraft had a 62 gal rear fuselage tank. Auxiliary torpedo drop tank 90 gal or alternative 180 gal tank. Late aircraft equipped to carry up to 1,000 lb bomb or two 300 lb RP load beneath each wing. Engine—Griffon 69 (max op 25 lb/sq in on 150-grade fuel, 18 psi when using 100 octane fuel).*

would allow for 5 mins full power take-off, climb to 20,000 ft with 15 mins for combat there and an hour's weak mixture flying, an additional five per cent taking account of likely engine development. Drop tanks of 50 or 90 gal would enhance reinforcement range.

In May the prototype was moved from Worthy Down to High Post and out of the Invasion area. Engine reduction gear ratios on Spitfire XIVs and 21s were reckoned responsible for too much tail vibration. Changes were therefore needed and also in the positioning of the pitot head. These factors would effect the Spiteful whose pitot head was then sited on the port wing leading edge. Contra-prop trials were continuing and in June 1944 serious trouble arose with lubrication. All flying was halted on June 24, except by propeller makers. When it resumed, transitional units had to be lubricated every ten hours. High gyroscopic loading couple on the two propeller shafts needed to be avoided to eliminate the possiblity of breakdown of the thrust bearing units. Flight testing for 100 hrs of a Rotol contra-prop on Spitfire *JF321* was ordered, its aerobatic performance being curbed. Contra-prop introduction was seriously delayed, and particularly on Spitfires. Not so, though, the decision to fit clear rear view teardrop canopies to Spitfire 21s. In June 1944 it was decided that the 101st *et seq* examples would have them, such aircraft becoming Spitfire 22s.

June 1944 found Supermarine still trying to get CRD agreement to Merlin-engined Spitefuls using the fourth as prototype. The MAP told them that it meant Rolls-Royce producing a special Merlin which, on May 29, that company has stated it was unable to produce. In reply Supermarine presented a paper proving a Merlin Spiteful's range was nine per cent better than a Griffon 65 version. Carrying 133 internal gal and a 90-gal drop tank, the action radius would be 300-325 miles and with a 200-gal drop tank about 700-760 miles. The CRD refused to budge — particularly as the F.1/43 programme steadily slipped. The MAP placed strong pressure on Supermarine to 'do something about it' and also to get the important Spitfire IX *ML186* with an extra 75-gal rear fuselage tank to Boscombe Down quickly. N.E. Rowe told Supermarine that he was 'very shocked' when, by June 1944, the prototype F.1/43 still had not flown. The reason, he was told, was that a contra-prop had not become available when expected, causing a switch to use of a five-bladed propeller. Rowe did not accept that and ordered the company to have the second F.1/43 flying within two months at the latest. Perhaps it was to calm matters that Supermarine agreed to forget the Merlin version. The interim Spiteful prototype *NN660*, comprising a Type 371 wing attached to a Spitfire XIV's fuselage featuring a production tail, made its first flight on June 30 1944.

Hawker by this time had two prototype Tempest IIs flying and a third, *JN750*, was almost ready to fly. Prospects remained poor because of the excessive engine vibration transference to the aircraft structure. To help cure this the first six production Mk IIs would also be utilised. Two F.2/43 Centaurus XII prototypes had been forecast to fly in June and July 1944. They would be late, and the Griffon examples had barely been started.

Between May 1 and June 30 1944 A&AEE tested Tempest II *LA602*. Its Centaurus V fitted with Siamese twin exhausts drove a 12 ft 9 in four-bladed propeller. Flying at 11,000 lb, *LA602* reached 440 mph at 13,700 ft using FS gear and 20,000 ft in 5.7 mins. Despite oil-cooling problems the service ceiling was 38,200 ft. A creditable climb to 10,000 ft in 2.4 mins was matched by rates of 2,850 fpm at 15,600 ft and 3,220 fpm at 14,900 ft for combat. Aileron control

— heavy at all speeds — meant adding spring tabs. Strong vibration was acceptable in early production aircraft, if alleviation soon followed. Starboard wing dip during a landing stall and with the aircraft too high was a dangerous feature for any inexperienced pilot. Insufficient elevator control prevented the tail coming down when landing engine-off. Clearly there was need for more pre-service development, serious because the Tempest II — and F.1/43 — were much needed in the Far East to replace Hurricanes and Thunderbolts. Delay was unfortunate because Centaurus engines were now becoming available, that for the sixth production Tempest II reaching Hawker on June 9.

Hearing claims that the Centaurus XIISM could boost the Tempest's speed to 470 mph at 25,000 ft, it was agreed that the fifth and sixth aircraft should become test beds. The engine, though, would not be available until summer 1945 so the Tempest II remained earmarked for the Centaurus V and possibly VII. Centaurus production at Hawthorn in 1943 was reduced, allowing increased Hercules output. No front-line aircraft then nearing service needed the Centaurus, but soon the extensive Buckingham bomber programme, the Firebrand, Brigand, Warwick II and Tempest II all made demands upon its production. A new Tempest II line was established at Bristol's Banwell plant and, with the Sabre still doing poorly, plans of July 1944 called for production of only Centaurus Tempests from June 1945. Peak production plans called for 230 Centaurus engines monthly; an impossible total. Sabre Tempests, it was therefore decided, would continue and constitute half the production. When Hercules output tailed off in April 1945, monthly Centaurus production was of about 140 engines; eventually it became 170.

Britain was unprepared for the need for fighters to engage fast, low-flying jet-propelled V-1 cruise missiles. Details of their deployment were known, but little about performance. The first salvo came from France early on June 16, and two Tempest V squadrons of 150 Wing, Newchurch, flew their first 'anti-diver' patrols soon after dawn. By the end of that day they had claimed eight flying bombs. These very small targets generally arrived at between 1,500 and 2,500 ft and at about 340-370 mph. Standing patrols needed to be flown to combat them. The only aircraft with sufficient speed and suitable for such daylight operations were Tempests, Griffon Spitfires and Mustangs. All soon had their engines uprated, changes to the Spitfire XIV including its acceptability of 150 octane fuel and +25 lb boost, adding 30 mph to its low-level speed. The small target, difficult to engage, meant a special RAE gunsight had to be fitted to 501 Squadron's night-fighting Tempests.

Many fighters became involved as the V-1 onslaught increased in ferocity until, on September 5 1944, the major phase ended, fighters were credited with 1,900 V-1s. By types their scores were: Tempest V 663, Mosquito (night) 486, Griffon Spitfire 342½, Mustang 247½, Spitfire IX 98½, Typhoon 30, Spitfire V 19, Meteor 1 8½, P-61 5. Top scoring squadrons were No 3 (Tempest) 257½, No 486 (Tempest) 221½ and No 91 (Spitfire XIV) 150.

For Hawker, July 1944 was a time of some sadness because Hurricane production ended. The Kingston works had produced 10,030 Gloster 2,750, Austin Motors 300 and the Canadians 1,451 — total, 14,531 aircraft. But Hawker's future was bright.

Chapter 9

Fast and furious

September 1 1944 witnessed the first flight of the F.2/43 *NX798* (Centaurus XII). P.G. Lucas after landing reported that the aircraft 'felt good'. Its stability was fine, ailerons effective. Take-off weight with 170 gal aboard was 11,150 lb — close to the Tempest II's — but take-off swing was reduced and the F.2/43 felt lighter on the controls than its predecessors. Finer proportioning, frictionless controls and spring tabs on both ailerons and rudder contributed to impressive qualities.

Widespread acclaim was heaped upon the new Hawker, N.E. Rowe and Camm favouring a score of examples with powerful Sabres easing demands upon Centaurus production. After 21 hrs flying achieved in 33 flights, *NX798* flew to the A&AEE. Their verdict was less enthusiastic than expected. Wing drop near the stall and at high speed was more troublesome than initially realised. Certainly *NX798* was superior in every respect to the Tempest and its speed reached 470 mph ASI. The newest Tempest V could reach 432 mph at 18,500 ft (FS) and 412 mph at 6,500 ft (MS), making it about 5 mph faster than the production Tempest II. Prior to SEAC service, trials of the latter would be undertaken by a squadron in Britain, then six Mk IIs would be tried overseas.

Hawker's F.2/43 was clearly successful and would probably be faster if

The Fury prototype, NX798.

Whilst the RAF never operated the Fury, the Royal Navy found the Sea Fury an excellent fighter-bomber.

Sabre-powered. Nevertheless, September 17 1944 brought an order for a replacement jet design to commence in early 1945 at Hawker. Already preliminary ideas for an advanced Gloster jet fighter were being explored, also a Jet Spiteful, a none too attractive alternative. Meanwhile the Spiteful's future was increasingly doubtful. Its top speed was disappointing and its engine ran rough, transmitting vibration throughout the airframe. A flexible engine mounting was necessary in the second prototype. The five-bladed propeller was also unsatisfactory and promoted aerodynamic imbalance and more requests for contra-props. It was believed that it was incompatible with the fuselage shape and that air flow from twin radiators brought additional vibration. On September 13 the prototype crashed, the cause never fully ascertained. Among

Tempest V SN219 spews smoke from its Smoke Curtain Installation gear during testing over Salisbury Plain (RAF Museum 6065-1).

A fearsome sight, and armed to the teeth, Tempest V JN740 carries eight 60 lb RPs for firing trials at A&AEE.

the foregoing problems a reason may have existed, but discoveries with the second prototype suggested problems with the flying controls.

September's plans were for two prototype F.1/43s, two navalised examples and a pre-production RAF machine. With the prototype lost, and naval variants overweight, prospects were not encouraging. Rotol devised a propeller to reduce vibration, and a four-bladed Dural propeller was fitted to discover whether vibration was due to propeller design or transmission problems. Now the Air Staff enquired whether the Spiteful's operational range could be increased, stating that extra tankage was necessary before Spitefuls entered Far East service.

The flying bomb campaign largely over, Tempest Vs were converted to fighter-bombers for 2 TAF. Late in October 1944 *MW735*, the first production Tempest II, first flew, two months late. Plans already called for F.2/43s to enter production in March 1946 and supersede the Tempest. Hawker, Bristol and Westland factories would between them deliver 209 Furies in 1946. Expected Tempest deliveries were 341 in 1944 and 555 in 1945.

Spiteful production was prescribed by the MAP on November 15 as: January 1945 — 1; February — 2; March — 3. Supermarine quickly pointed out that, due to the crash, the second prototype would not be flying before December 15 1944, and doubted whether production could commence before March 1945. The MAP asked for another review of the Spiteful's lateral stability in November, following one six months previously showing poor line characteristics. Snatching ailerons remained a problem needing much attention. *NN660* had shown regular violent oscillation and vicious stick movement close to the stall, making the Spiteful unacceptable without major modifications. New aileron control runs were necessary and had to be installed in the already plated wing of *NN664*, seriously delaying it. All had moved far from the original concept of laminar flow wings fitted to production Spitfire VIIIs in the belief that a speed of 493 mph at 24,500 ft was attainable if five-bladed or

A combination of rocket and cannon, four-bladed propeller and clear view canopy transformed the Typhoon into an excellent ground attack fighter (H.A. Kofoed).

contra-props were fitted. Interestingly, a modified Spiteful eventually reached 494 mph!

Vickers informed the MAP on December 16 1944 that Rolls-Royce had, on December 13 1944, notified the company that the second Spiteful's engine had suspect pistons which needed replacing. 'That', Joe Smith wrote, 'must delay the aircraft again.' Rowe, in his reply, stated 'Your letter of December 16 disappointed me. I expect news of the first flight before Xmas, as a Xmas present.' He had already told Smith, sharply, that he was to make it his 'personal job' to see that *NN664* flew soon. The first flight came on January 8 1945 and inverted aileron action, aileron snatching and wing dropping prior to stall were soon experienced. Discovery was made of the forward wing area stalling before the remainder.

If the Spiteful was not doing well, auguring ill for Supermarine's fighter future, the Spitfire displayed performance beyond any original concept, as these figures show.

Mark	Max speed (FS) (mph)	Rate of climb (fpm)	Service ceiling (feet)	All-up weight (lb)
I	362	2,530	31,900	5,820
Vc	367	3,850	37,500	6,750
VIII	410	3,750	42,000	7,800
XIV	446	3,600	40,000	8,400
21	460	5,000	43,000	9,000

Time taken to get Griffon Spitfires and the Spiteful into service was more a reflection of the persistent excellence and value of earlier Spitfires than serious problems with the later aircraft. At the end of 1944 production commenced of the Spitfire 22 featuring the new rear view cockpit but otherwise similar to the F 21. In December 1944 the teardrop canopy was introduced also to production Spitfire XVIs as Modification 963 and to Mk XIVs as Modification 1224. So successful had the interim Mk XIV proven that on December 29 1944 agreement was reached to produce a fighter reconnaissance version of the aircraft, the FR

Most difficult of all Griffon-engined Spitfires to develop, the Mk XXI is represented here by LA215 (Supermarine).

Mk XVIII. Spitfire 21 production was slow, the aircraft beset by handling problems and for almost a year unacceptable for squadron service. Delivery to maintenance units began in September 1944, but it was January 4 1945 when the first two examples were placed in 91 Squadron.

The question of fitting contra-props to advanced fighters remained unsettled. Spitfires *JF317, JF320* and *JF321* were testing Rotol types and Mk XIV *RB144* at Hatfield had a 10 ft 4 in type. An 11 ft Rotol was at High Post for Spitfire 21 *LA213* and two fully modified Griffons, in production state, had been built. Farnborough's December 1944 tests established the effects of multi-blade propellers on Spitfires, finding that a 60 per cent increase in tailplane area much

Sandwiched between a late Griffon-engined Spitfire and a Mk XIV, two hybrid Merlin-engined machines in a formation paying tribute to Group Captain Douglas Bader, at the 1983 Greenham Common Air Tattoo (Author).

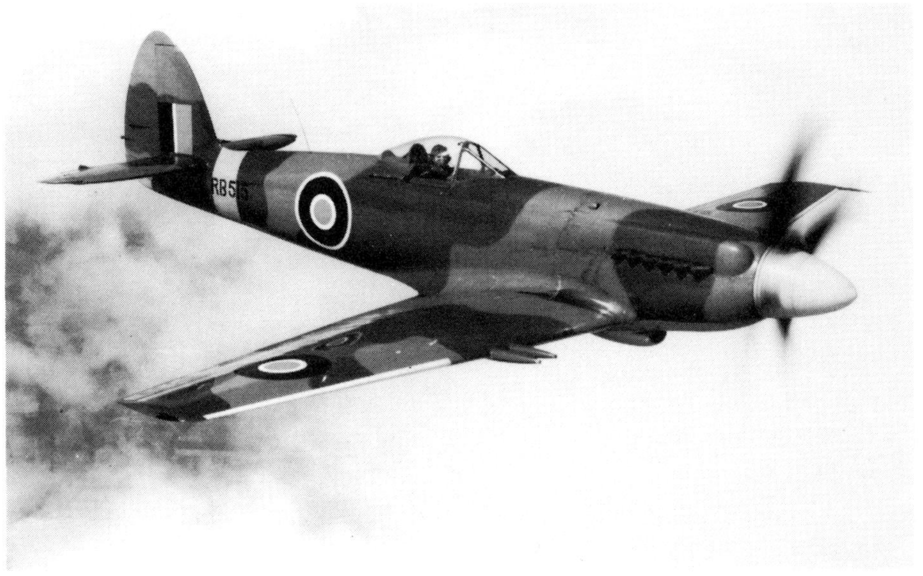

Contra-prop stable mate of Tornado R7936, *Spitfire XIV* RB144 *used at Hatfield on November 21 1944* (BAe).

improved tractability. Five blades showed a distinct destabilising effect, but not the contra-prop. The Spiteful was found to need a 100 per cent increase in tailplane area when fitted with a five-bladed propeller so, on December 18 1944, the MAP told Supermarine that all Spitefuls must have an enlarged tailplane, meaning yet more delay. Spitfire 22s with teardrop canopies and cut-away rear fuselages were also to have the new type of tail unit.

The stream of criticism directed at the Spiteful was demoralising to all

Spiteful XIV RB515 *displays a fine finish, essential in order to obtain most advantage from laminar flow wings* (RAF Museum 6068-1).

Spitfire FR XIV MV259. Clipped wing tips, teardrop canopy and camera port aft characterise a very efficient Spitfire (Supermarine).

concerned. Some proportion of it was undoubtedly the result of the company working on the transonic zone which brought many worries and problems. As Commander Bird of Supermarine wrote on January 7 1945, 'the Spiteful is the first thin wing aircraft, and must be regarded as a new design with teething troubles expected'. Indeed, it was still very experimental as it entered production, which meant the incorporation of modifications which could only bring about major problems. In view of these points another 150 Spitfires were ordered to bridge a possible gap antagonised by snaking problems afflicting the Mk 21 and soon the Mk 22.

Revision of the original Spitfire Mk XIV had led to the appearance in autumn 1944 of a fighter-bomber version and the Mk XIVe armed with two 20 mm guns and two .50 in machine-guns. Initial modification had resulted in the FR XIV differing from the XIVes modified retrospectively to FR (and officially called FR XIV) by having a 31-gal auxiliary fuel tank and rear view canopy. Estimated maximum take-off weight was 8,980 lb and landing weight 8,750 lb. Due to their

Spitfire XIVs saw considerable Far East service, among them NH804:O and TZ184:Z.

Superbly pugnacious, its engine almost audible as it formates, Spitfire FR XVIII TP448:GZ-P *of 32 Squadron* (RAF Museum P10503).

similarities the FR XIV (LR) was the production version, whereas FR XIV applied to converted aircraft. Both had .50 in guns, the 'e' suffix therefore being unnecessary. Only oblique cameras were carried — in the rear fuselage — whereas the Griffon 65 FR XVIII held two vertical cameras. Spitfire FR XIV(LR)s replaced 268 Squadron's Typhoon FR Is which, similarly, did not exist as 'FR Ib' since there was no FR Ia. Such precise designations were often never promulgated. In June 1945 Spitfire XIVs first reached the Far East, for 11 Squadron in India.

Variants of the Mk XVIII were the F XVIII with a 62-gal rear fuselage tank and no cameras (all-up weight 9,190 lb) and the FR XVIII with 31-gal rear fuselage tank and provision for vertical/oblique camera (all-up weight 9,000 lb). Mk XVIII production replaced that of the Mk XIV in March 1945, the first example, *NH872*, flying in June. A strengthened undercarriage, as fitted to late Mk XIVs, was essential for fighter-bomber versions of the Mk XVIII whose operational weight was above 10,000 lb.

Sufficient modified Spitfire 21s reached 91 Squadron at Ludham to allow operations to commence on April 10 1945. Five pairs of Spitfires flew anti-shipping sorties during the day, *LA203* and *LA229* being shot down by flak. On April 26 *LA252* and *LA223* destroyed a midget submarine. Such operations, and those connected with V-2 operations, totalled 152 sorties when operations ceased on May 1, *LA206* and *LA197* flying the final sorties.

In June 1945 No 1 Squadron became the second to have Mk 21s. February 1946 saw a third squadron, No 122, begin briefly to use them. It was April 1946 when the only other Mk 21 Regular Squadron, No 41, was equipped.

The intention was to form two Tempest II squadrons in March 1945. Instead, a handful of Tempests joined 13 OTU at Finmere. The first production Tempest II was delivered on November 9 1944 for trials, and on March 8 1945 the first Service example. A forecast of April 20 1945 indicated that between April and December 1945, 423 Mk IIs would be produced and a further 213 in 1946. During July-August 1945 Tempest IIs at last entered squadron service, equipping 183 at Chilbolton. Next month 247 received some, and 183 Squadron

Adding a contra-prop to Mk XXI LA215 entailed a modified engine. Note the different exhaust stack (RAF Museum P13566).

was renumbered 54 Squadron in November. Revised plans of June 6 1945 had called for 390 Tempest IIs to be delivered between February and May 1946, for which purpose Fury production was retarded by six months and programmed to commence in August 1946, allowing for 22 examples to be delivered by December 1946 instead of the 103 previously expected. Late in August 1945 Tempest II production plans were adjusted again, the monthly 30 to be reduced to 25 from June 1946.

Summer 1945 saw trials of contra-props continuing, Spitfire 21 *LA213* having a 10 ft 4 in production type Rotoloid wooden-protected contra-prop and a large compensating fabric-covered rudder, testing having started in March 1945. De Havilland was trying an 11 ft contra-prop reckoned best for the Spitfire 22. With the war ending there were many surplus propellers, an aspect unfavourable to contra-prop production. Nevertheless, test flying continued

To assist development of the sophisticated Seafire F 47, Spitfire Mk 22 PK684 with some modifications bringing it into line with a Mk 24 has a contra-prop (RAF Museum P9171).

Spiteful prototype NN664, *unpainted, made an unexpected appearance at the Marham flood relief display in 1947* (Author).

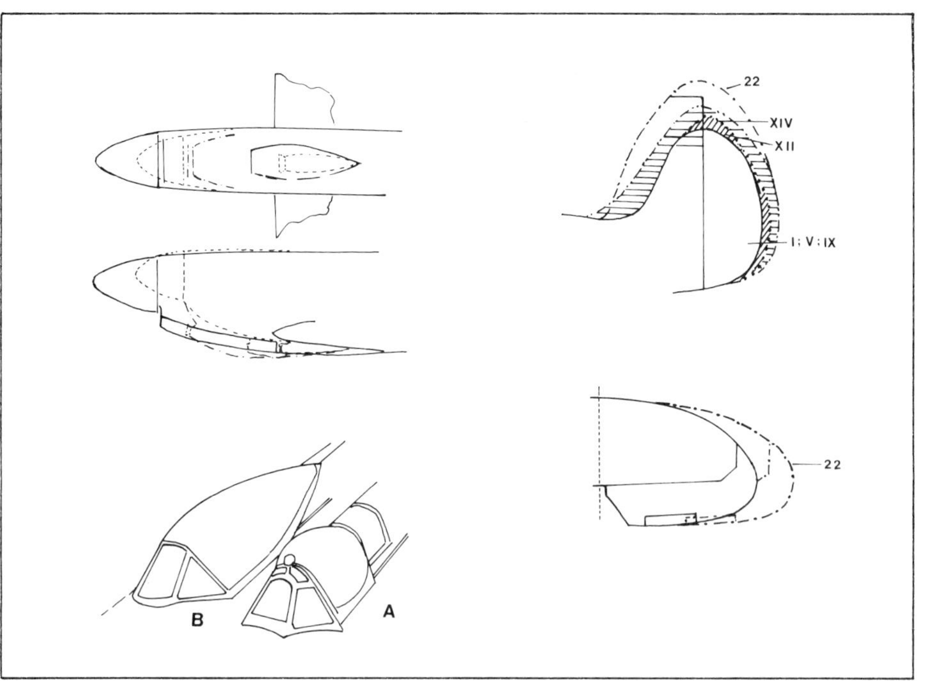

Above right *Outlines of Spitfire tail development. Maintenance of reasonable fore-and-aft stability was a problem throughout the Spitfire's development. Every small modification affected this. Corrections included increased area elevator horns, inertia weights in elevator controls, changes in elevator section, metal skinning of elevators. More powerful engines, higher solidity airscrews, additional fuel loads and changes in operational altitudes also affected directional control. This brought necessary changes in fin shapes, rudder shape, rudder horns, anti-balance tabs, metal-skinned rudders. Changes applicable to various marks of Spitfire are indicated.*

Above left *Alterations to Spitfire nose profile and varying air intakes. Lengthening of the aircraft's nose, a variety of air intakes to cope with various situations and more powerful engines were all Spitfire features. To counteract the destabilising effects of longer noses, increased vertical area aft had to be fitted.*

Below left *Improved view aft was constantly sought for all fighter pilots. Clear view panels, stronger windscreens, fitting of mirrors, bulged side and roof panels and ultimately one-piece blown hoods aft of narrow framed windscreens were designed for Spitfires.*

until February 1946 when plans for their use on the Spitfire 22 were cancelled. Available contra-props would be fitted to Seafires, experimental aircraft and later to Shackletons.

Before becoming acceptable for squadron service, Spitfire 22s required extensive alterations including Modification 1613, the fitting of a new tail unit first flown on Spiteful *NN664* on June 24 1945 and ordered to be standard on all 278 Spitfire 22s. Applied to the Spiteful it was not the prophesied cure-all, inducing wing stalling and requiring an increase in wing dihedral on all Spitefuls and leading to modifications of wing leading edges. Although Spitfire 21s and 22s were cleared to carry external loads of up to 2,000 lb, they served both as interceptor and ground support aircraft.

With jet fighters scheduled to take over high altitude defence but lacking in low level endurance, the Air Ministry informed Supermarine on August 23 1945 that the Spiteful was to be dropped from the fighter programme other than as a fighter-bomber and, unless cleared for this role by mid-1946, would be dropped entirely. Its future in the fighter policy review of June 6 had been bleak. With delivery to commence in October, only 13 Spitefuls could be expected in 1945, a far cry from the 88 planned in April, and 248 in 1946. These policy changes led to official vacillation on the aircraft's role and its nomenclature was curious. The first production examples were Mk XIVs, this designation stemming from their affinity to the Spitfire XIV and leaving lower mark numbers free for Merlin Spitefuls. A typical Mk XIV had a top speed of 483 mph at 21,000 ft. No contra-prop Mk XVs were built, and eventually only 17 production Spitefuls, produced between February and December 1945. *RB518* had, instead of the usual 2,375 hp Griffon 69, a three-speed Griffon 101. With +25 lb boost, it managed an amazing 494 mph at 27,800 ft. Later, it mounted a Griffon 121 which drove a contra-prop. The aircraft, known as the Mk XVI, was damaged in a crash landing.

Quite unlike the Spitfire, in appearance and much else, development Spiteful XIV RB518. *The wide track undercarriage is a very noticeable feature* (RAF Museum P5907).

Angular outline laminar flow wings to make production easier are featured by Spiteful XIV RB515 (RAF Museum 6068-15).

A ground-attack role in the coming jet age was the Spiteful's only possible future so *RB516*, suitably kitted out, was ready in April 1946 for A&AEE trials and joined *RB517* undergoing engine cooling and performance testing. *RB523*'s armament installation was being cleared, whilst *RB518* was used for propeller development and, with a Griffon RG35ML, was then destined for its contra-prop future. *RB519, RB520* and *RB521* — all non-standard — were used for general development. Chord added to the wing leading edge had, according to Smith, improved control down to stalling speed.

After three year's effort, the Spiteful's future faded with alacrity following a damning A&AEE report. *NN664* it described as 'generally below average', with a poor cockpit layout and too lengthy operational turn-round between sorties. Poor equipment layout, trouble in keeping the aircraft clean, bad skin plating technique, difficult access to armament, badly fitting panels and many more criticisms damaged the Spiteful's already dented image. Supermarine's first jet fighter, the E.10/44 Jet Spiteful, would soon be flying, but already it was being judged 'inferior' because of the Spiteful's reputation. Much more was at stake for Supermarine since, with unit costs rising fast, money short and the RAF's future uneasy, spending on military aircraft must be less. Quite abruptly Super-marine, the company which had given Britain both the Spitfire and a precious symbol of victory in virtual defeat, suffered decline. Its manner was not dis-similar to that of the nation's great wartime leader. Competition would long continue, but there was room this time for only one winner, ultimately, and that came from the favoured Hawker kingdom.

Chapter 10

To keep the peace

Hawker production centred upon the Tempest II, 50 of which were built by VE-Day and 183 delivered by VJ-Day. The Centaurus Fury F1/P1, was coming along well, its specification agreed on May 13 1945. *NX798* performed well and 15 pre-production development Furies had been ordered. One of the RAF's five would have a Sabre VII because on June 16 1945 MAP asked the Air Staff whether they would accept Sabre Furies, releasing Centaurus engines for Sea Furies. One of the RAF's five Furies, *LA610*, first flown on November 27, was Griffon 85-powered and fitted with a contra-prop. It attained over 480 mph, which was encouraging. The RAF's second prototype, *NX802* (Centaurus XII), first flew on July 25 1945, but since March 13 priority had switched to naval variants, delivery of which was planned to start in January 1946.

There was another aspect to Fury production which was linked to the Sabre. To keep in business Napier needed to deliver a dozen Sabres a month, perhaps until 1949 when their revolutionary Nomad compound-engine should enter production. That would mean that a third of the RAF's reciprocating fighters would be Sabre-powered, which was quite unacceptable. Plan C called for a monthly production of 14 single-engined fighter-bombers with reciprocating engines, whereas existing production plans for the period January 1946 to April 1947 had been set in 1945. For Spitfire F XVIII — 14; Spitfire FR XVIII — 13; Spitfire 22 — 137; Spiteful — 43; Tempest II — 129 and Tempest VI/Fury — 173. Of these the Tempest II was the only specialised ground-attack aircraft, but the Air Staff were keen for Supermarine to continue supplying the RAF as a back-up source to Hawker. Supermarine's position seemed likely to be helped by a Canadian order for 139 Spitfire 22s, but it never materialised. To keep the firm alive the RAF opted for five Spitfires a month in 1946.

Pressure upon the RAF to adopt the Sabre Fury continued and resulted in *LA610* being fitted with a Sabre VII for trials. The RAF, though, took a strong stand and wanted the Fury cancelled or developed only for the Navy and certainly out of production in 1947. Much preferred was the Tempest VI, Sabre VII-powered. Since the Navy had no use for the Sabre Fury its development for the Services was halted on August 28 1946, although Hawker maintained the programme against possible overseas sales. Napier would have to get by building Sabres for Tempest VIs.

Tempest IIs (Centaurus V and normal fuel tankage 158 gal) were progressively modified on production lines to the 130th aircraft. *MW801* delivered in December 1945 was a typical tropicalised example. With a tare weight of 9,300

Above *Production Tempest FB IIs like* MW742 *initially carried very distinctive white bands on nose and tail, ensuring that they were not misidentified* (BAe).

Below *Graph of the Tempest's performance, speed and climb.*

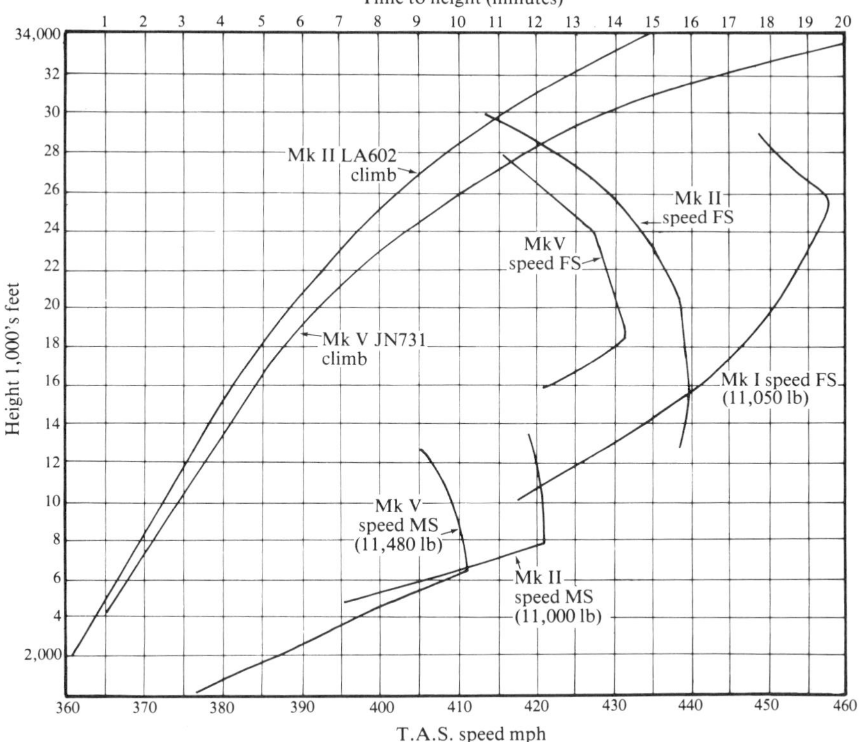

lb and weighing 11,824 lb with two 45-gal drop tanks fitted, its top speed was 442 mph at 15,100 ft. It was typically a Tempest II in another respect, in that it eventually made its way to the Indian Air Force like many others of its breed. Tempest II production far exceeded British needs.

With Sabre V reliability sufficient, a 100-mile range extension, 18 mph speed increase and faster climb were forecast from its wedding to the Tempest VI, which was given the go-ahead on June 6 1945 when an order was placed for 300 for ground attack duty overseas. A 2,340 hp Sabre V was initially tried in *HM595* and first flown on May 9 1944. Engine cooling was a problem since the radiator occupied the entire nose bath space causing the vital oil cooler to be resited. The only possible position was in the wing root where carburettor intakes were already fitted. Less vulnerability to ground fire remained an advantage held by the Tempest II, production of which was ample and overseas clearance near. Favour soon reverted to the Mk VI, however, for Napier's Nomad compound-engine for long-range aircraft could only be developed if the company was kept alive on Sabre production, further reflected in the Sabre Fury programme. By autumn 1945 it was agreed that Tempest VI production would begin in February 1946 and, under Plan Stage 3 Post-War Air Force, it would Mustangs in the Middle East.

Although the first production Tempest VI (weight, loaded with eight RPs, 13,201 lb) was delivered in November 1945, Ministry dissatisfaction with engine cooling remained. Two oil coolers were now fitted, but it seemed that a full cure could take a year and severe financial cut-backs made the whole venture questionable. That secondary oil cooler had not been initially envisaged, and it reduced the wing fuel load by 26 gal, wiping out any range advantage over the Tempest II which regained its popularity. Hawker, shaken by the strong Ministerial stand, worked intensively and the aircraft passed its tropical trials at Khartoum in December 1945. Production was cut to 142 Mk VIs, ended late 1946, and the aircraft served in Germany and the Middle East.

Improved reliability, but high fuel consumption at low and medium levels, meant that jet fighters for the forseeable future must handle only the high level interception task. Piston-engined fighters would be given the ground-attack role. On November 20 1945 major policy decisions affecting Fighter Command's future were taken. Tempest IIs in a trials squadron would be replaced by Vampire jets in March 1946, Vampire problems by then hopefully cured. A second squadron would re-equip in June 1946. Remaining Spitfire IXs would become quickly available for foreign purchase in an attempt to arouse interest in British products. Spitfire XVIs (Packard Merlin 66), production of which ceased in August 1945 (by which time over 7,000 Mk IX/XVIs had been produced), would replace them in four squadrons. Spiteful production (eventually totalling 17 and two prototypes ended in December 1945) was briefly re-instated, one squadron to have them for trials in place of Spitfire 22s. Protracted Spiteful ground-attack role development halted this idea and, in any case, the general opinion was that the Spiteful was little better than the troublesome Spitfire 22. An additional Regular ground attack Spitfire 21 Squadron would form in December 1945 because of Mk 22 delays. Fighter Command day fighter strength would be 20 squadrons in December 1945 and rise to 22 late in 1947. Strength of each would be 16 UE — 18 UE on Auxiliary Air Force squadrons of which 6½, using Spitfire 22s, were scheduled to form by June 30 1946. The last Spitfire 21 delivery came in January 1946, the type's

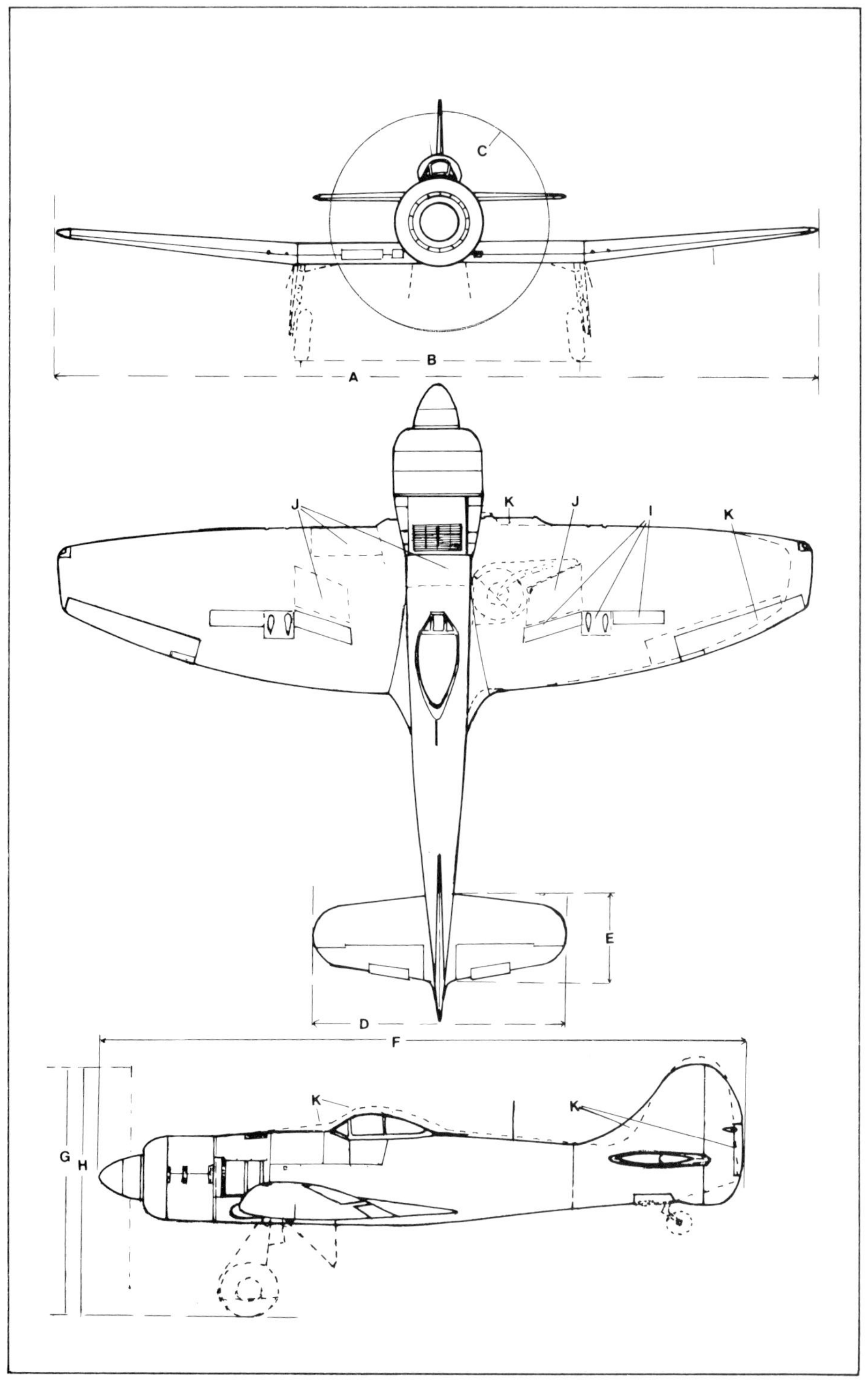
C
A
B
J
K
J
I
K
E
D
F
G
H
K
K

Teardrop-canopied Spitfire F 22 PK312 leads two Mk XXIs LA217 and LA232. The Mk 22 had a lengthy development period due to its being switched from interceptor to ground attack role (RAF Museum 6044-9).

surfeit being so high that three AAF squadrons had them instead of Mk 22s.

The 236 Spitfire 22s delivered between March 17 1945 and February 11 1946 were mostly earmarked for service as ground-attack aircraft overseas, after Fighter Command had assessed the type in service. Not until December 1945, however, was a Mk 22 modified sufficiently for trials, and these revealed unacceptable vibration. Although bombs and drop tanks were intended for the Mk 22, the aircraft was never cleared to carry them. Such items were a feature of the refinement, the Mk 24, the first of which was delivered on April 13 1946. A month later operational standards of AAF squadrons were ordered to equate those of the Regulars. All day auxiliary fighter squadrons would use Spitfire 22s

Opposite *The Hawker Tempest II; Fury wing and fuselage shown for comparison. Key: A — wing span 41 ft 0 in (root aerofoil H.14/14/37.5, tip H.14/10/37.5), B — 14 ft 10½ in, C — 12 ft 9 in, D — 13 ft 9 in, E — 4 ft 8 in, F — 34 ft 5 in, G — height, tail up, one blade vertical, 13 ft 6 in, H — height, tail down, one blade vertical, 14 ft 6 in, I — covers over ammunition boxes, supplying four Mk V Hispano 20 mm cannon, 162 rounds to inboard guns, 156 rounds to outer guns, J — fuel tanks, main tank 76 gal, each insterspar tank 28 gal, nose tank 26 gal = 158 gal fixed tankage, K — reduced span (38 ft 4¾ in) of Fury, also length 34 ft 3 in. Tempest II gross wing area 303.7 sq ft: inner wing dihedral nil, outer 5° 30'.*

Tempest IIs carrying bombs or rockets saw post-war service overseas prior to delivery of the Vampire FB 5. PR632:RS-A is seen here in India, in 30 Squadron's hands.

in an interceptor role and have Mk 24s later, following release to service of the Mk 22 on May 1 1946. Some months would elapse before the squadrons were fit to train in an RP/FB attack role. A dozen AAF squadrons eventually operated Spitfire 22s, as interceptors only.

Following clearance for overseas service, Tempest IIs joined 5 Squadron in the Far East in March 1946 and later served with three other squadrons. The quantity produced, 482 when the last built, *PR921*, was delivered on June 10 1946, could never be fully utilised. A further use for them had been decided shortly after the war when a decision was taken to arm seven BAFO ground attack squadrons with Tempest IIs at 16 UE. By early 1946 this, for financial

Post-war fighter-bomber/reconnaissance aircraft types used in the Near and Middle East. A Mosquito FB VI (TE912) leads a Spitfire FR XVIII (TP264) and three Tempest VIs (NX182 nearest) (RAF Museum P014945).

Rare in all senses, a Spitfire F 22 PK674:L *in front line squadron service, with 73 Squadron, the only Regular Mk 22 squadron. Photographed in Malta, the aircraft in post-war type roundels carries the squadron's blue and yellow colours as shown in Chapter 3* (RAF Museum P13691).

reasons, had been cut to three squadrons. No 26 received its quota mid-1946, No 16 in summer and No 33 in autumn 1946, all as Tempest V replacements. Tempest IIs served 16 Squadron until December 1948 and No 26 until April 1949. In July 1949 No 33 Squadron took its Mk IIs to the Far East, operating them during the Malaysian emergency until late 1950.

Postwar Fighter Command strength under 'Plan D Target Force' meant considerable cuts but, on August 29 1946, it was decided not to act on them until immediately after the Battle of Britain flypast, then reducing the Command to an equivalent of three night fighter and nine day fighter squadrons by disbanding 151 Squadron and halving the strengths of the existing six night fighter and 18 day fighter squadrons. Esprit de corps and squadron identity were thus retained, re-equipment would be accelerated and exercises continue at squadron level. Aircraft types, uncertain, would depend particularly upon airfield availability. Regular day fighter squadrons needed jet fighters, with Mosquitoes, although obsolescent, arming night fighter squadrons. When production and training permitted, the Auxiliaries would be similarly equipped.

The first Spitfire FR XVIIIs, for the Middle East, were despatched in June 1946 to replace F XVIIIs declared obsolete on May 17 1946. Mk XVIII production ended early in 1946 but not until January 1947 did No 60 Squadron receive, in Malaya, its quota of the 300 built. Already the Spitfire and Tempest were outdated support fighters, however, and unable to protect themselves. So in February 1947 the Nene-powered Vampire 4 jet fighter became scheduled to provide protective top cover to the piston-engined aircraft.

In August 1947 the last front-line Spitfire 21s retired from 41 Squadron, leaving only three AAF squadrons using that type, Nos 600, 602 and 615. No 73 Squadron at Takali, Malta, was from November 1947 to October 1948 the only regular squadron ever to have Mk 22s. Supermarine delivered their last Spitfire on February 20 1948, a Mk 24, *VN496.* Fitted with two 33-gal rear fuselage fuel tanks, only 54 Mk 24s were built, and 27 Mk 22s were converted into Mk 24s. Only No 80 Squadron used that type, between January 1948 and January 1952.

Memories of amazing advances in the 1940s are awakened whenever a remaining, airworthy Spitfire such as AR213 appears. This hybridised Mk Ia is powered by whichever Merlin variant is available, and uncharacteristically has a four-bladed propeller.

Initially the squadron was based at Gütersloh, Germany, but in July 1949 moved to Hong Kong for dual air defence/ground support duty.

The last front-line Spitfire fighter was withdrawn from squadron service in January 1952, whereas the last front-line Hurricane had retired in September 1945. Spitfire dominance of the fighter field is obvious, and it held its own well into the jet era. Throughout the interceptor story the Spitfire's excellence can be seen as due largely to improved engines and armament — particularly the addition of heavy cannon, the introduction of which, to the British fighter scene, is related in Part Two.

Part Two:

The cannon fighter

Production Whirlwind P7048 was delivered on May 27 1941 and served with 137 Squadron from November 27 1941 until May 31 1943. It eventually became G-AGOI (RAF Museum 6013-12).

Chapter 11

Big guns

Cannon were the primary armament of fighters for most of the war. Perhaps surprisingly, such guns had long been considered to require special airframes, but this soon proved not to be the case. Fast fighters needed to deliver decisive punches in ever briefer engagements. Destruction of metal structures needed the use of heavy calibre shells; cannon becoming feasible when all-metal structures became available to absorb recoil forces. Such guns were not new, Westland in 1927 installing a 37 mm cannon in a fighter. Bristol and Vickers had similar interests but smaller-calibre weapons remained acceptable until protective armour was usual in aircraft.

On June 21 1935 the F.10/35 'fighter for the 1940s' was cancelled because 1934's designs revealed excellent potential. A superior fighter, for 1938 production, was to be devised and feature a flexibly mounted Coventry Ordnance Works (COW) 37 mm gun operated by a second crew member. Such an idea held sway until August 1935 when Squadron Leader R.S. Sorley revised F.10/35, concluding that multi-cannon armament should be featured. This was a turning point; an aircraft was to be designed around its weapons system. Four fixed 20 mm or 23 mm 60-round wing-mounted guns firing outside the propeller arc were suggested, despite possible stoppage problems and others associated with wing guns. It was thought that cannon-bearing wings might be devised for the F.5/34, perhaps the Hurricane and Spitfire, four cannon weighing little more than eight machine-guns and their 2,400 rounds.

Foreign designs had a cannon, mounted on or within the engine, but the chance of success using one gun seemed slight. At least four were needed, as required by the revised F.10/35, as well as a top speed of 330 mph at 15,000 ft, which was recommended to the CAS on November 11 1935 and agreed on November 30. The problem was where could the guns be obtained? Aero Engines Ltd, of Bristol, held a licence to manufacture French Hispano Suiza engines and cannon within the British Empire so, in December, the company was persuaded to send a representative to Paris to acquire cannon for test purposes. It transpired that privately developed 20 mm guns could be delivered by spring 1936, but not Hispano's new 23 mm gun developed for the French Air Ministry. Aero Engines pointed to the difficulty of installing bulky guns in fighter wings, adding that much development had been done on 'moteur cannon'. An example of such an engine which they were acquiring gave 956 hp at 11,500 ft. But the Air Ministry was interested in cannon only and enquired about costs and production rates under licence. Purchased directly from France, each gun would cost about £1,300, the company being prepared to supply two cannon in three months and another six in six months — sufficient for two prototype aircraft.

Redraft of F.10/35 became the F.37/35 which the Air Staff decided should be a custom-built cannon fighter. It would be single-engined, operate by day and night and be able with four cannon to obtain a decisive result from longer range than a machine-gun fighter. An engine-mounted gun seemed unlikely to have enough success. The fighter must be 40 mph faster than contemporary bombers, and have a 30,000 ft ceiling. Climb rate would be secondary to speed and hitting power. Allowance for increased engine power during the type's life span and possible additional tankage were required, and a take-off run of 600 yds — 700 yds if really necessary. Four wing cannon could be supplemented by moveable guns controlled by the pilot who needed a good view from his enclosed cockpit. Other features prescribed were easy ground accessibility to guns, retractable undercarriage, wheel brakes and electric starters — a number of novel features. On April 4 1936 all ideas of traversing guns, including cannon, were abandoned.

Armstrong Whitworth, Boulton & Paul, Bristol, Fairey, Hawker, Supermarine and Westland were invited to tender, and the Air Ministry ordered the Hispano Suiza Type S cannon along with a Dewoitine 510 moteur cannon fighter for investigation. All firms able to tackle F.37/35 had work in hand, however. Modifications were made by their builders to the Hurricane and Spitfire layouts, but conversion was more difficult than expected. Drag from the installations seemed likely to reduce performance considerably, this confirming to some extent the need for a specialised machine. Westland, who had built such aircraft, had spare design capacity. Fairey, with 20 mm cannon experience in the Fantôme biplane, were not yet fully committed but had naval interests. Supermarine were over-worked and Armstrong Whitworth, reckoned as skilled in weaponry, were producing bombers.

Westland it must be, a firm floated as a subsidiary of Petter's in 1935 following the success of the Lysander. Petter, a family engineering firm which knew little government encouragement having unsuccessfully tendered to both 1934 fighter specifications, found itself in poor health and survived by building Hawker Hectors. The workforce, reduced to 300, built stainless-steel beer barrels sold at a loss by Mr Petter as the Lysander was devised. Love of aeroplanes kept together Messrs Davenport, Penrose and Petter; the Lysander was their life line. Westland viewed F.37/35 with considerable, careful interest. If, for example, the radiator could be unobtrusively sited in the wing, much drag could be avoided. This suggested twin-engined layout was likely to face general opposition and Rolls-Royce's known mistrust of buried radiators. Although less manoeuvrable, it was a twin-engined design which Petter presented to the Air Ministry in February 1936. He enjoyed an encouraging response.

Supermarine, whose four-cannon Spitfire was unacceptable, realised the future lay with cannon fighters. Specification amendments permitted a twin-engined design, Supermarine tendered such a layout. Hawker in April 1936 submitted plans for a cannon-armed Hurricane, its likely speed only 270 mph. Bristol sent plans of their twin-engined Type 313, and Boulton & Paul's entry was powered by a Vulture.

At the September 1936 Design Conference, Westland quoted £27,500 as their prototype's cost and £18,000 for a second; Boulton & Paul cited £20,500 and

£17,500 and Supermarine £22,000 for one aircraft. Order of preference for single-engined designs was Boulton & Paul, Bristol, Supermarine then Hawker. Supermarine's twin was favoured because of the Spitfire's quality, although its 27-month prototype delivery date seemed excessive. Westland's design ran it a close second, the under-utilised firm deserving reward after its successful, difficult Lysander design.

Novelty hallmarked Westland's submission, with its thick magnesium alloy-skinned monocoque fuselage making it lighter than its aluminium counterparts. Wing leading edge, drag-reducing radiators ensured less engine frontal area. Internal exhaust ducting was planned while fuel tanks were integral with wing structure, and Fowler flaps would be used for the first time in Britain. A special Westland device would reduce wing area, outer wings incorporating broad chord lifting slots.

Provisional orders were placed for two each of Westland's and Boulton & Paul's designs, and for one of Supermarine's. Like Westland, Supermarine spent October 1936 moulding projects around the modified Rolls-Royce Kestrel engine, the KV26 Peregrine. Some officials maintained that accuracy and hitting power would be lost by not fitting moteur cannon in this engine, and by adding only 150 lb armament might thereby increase to six cannon in a twin-engined layout. Moteur cannon were considered by the Air Staff to be too complex, however, and it was also believed they would extend the five years taken to develop an engine, that power plant removal would be difficult and low drag cowlings and cooling systems spoilt. Rolls-Royce were prepared to develop a moteur cannon — but only if the Air Staff required it. They did not.

Westland faced two basic problems. Wing-mounted cannon required strong, heavy structures — performance influencing and demanding powerful engines which in turn needed plentiful fuel. Alternatively, utilising fuselage cannon confirmed that a radical two-motor layout with nose guns was easily accessible.

Merlins were ideal for a twin-engined fighter. For a heavy, single-engined machine there was no power plant powerful enough. Therefore the cannon fighter must be a 'twin'. But Westland, by opting to keep the aircraft as small as possible and to utilise light Peregrines cleanly cowled, were making unknowingly a fundamental error.

The Air Ministry confirmed the order for two trend-setting Westland P 9s on February 11 1937, cancelling all others.

Delivery of two prototypes, four cannon closely grouped in their noses, was needed between August 1938 and February 1939. Twin fins and rudders were replaced during project definition, by a single fin and rudder lifting the tailplane above fast airflow streaming from engine nacelles.

Prototype construction commenced in May 1937. By June the engine mountings and tailplane were being built, and work started on the forward fuselage in July. Main P 9 component assembly began in December 1937, construction of the second prototype starting in January 1938. On March 9, the promised engine not having been delivered, the workforce was switched to the second prototype upon which it was engaged until an engine arrived — in June. A 12-gun mock-up machine-gun nose was built in case cannon problems arose and in July it was decided to fit Exactor throttle controls.

Complicated and novel features, a three-month undercarriage delay, late delivery of oil coolers, component installation problems, failure of Hindu-minium wing castings and faulty fuel tanks all further delayed the aircraft.

Westland, trying to hasten things, attempted to secure a production order. Unimpressed by the firm's performance so far the Air Ministry refused to be committed until the prototype had proven itself, adding that unless the prototype was ready by December 1938 the programme would be ended. That meant the company working 24-hr days to meet the deadline. It was not all Westland's fault and although they were aware of being committed to a very ambitious project they forged ahead planning to produce F.37/35s between June 1940 and March 1941. Air Marshal W. Sholto Douglas, ACAST, thought this a very late time scale for a 1936 design, and private estimates by Mr W.E. Petter, Westland's Chief Designer, were even more pessimistic. Therefore the company was invited to pass production to another contractor. Fairey were proposed; Hawker showed interest as production at Yeovil was reckoned impossible in the ill-equipped workshop and by a firm which could not increase its labour force prior to firm contract.

Supreme, creditable effort resulted in taxying trials of the P 9 on October 5 1938. Quickly dismantled, it left by lorry on October 8 for Boscombe Down. There, flown by Harald Penrose, Westland's Chief Test Pilot, the all-grey prototype *L6844* took off at 5 pm on October 11 for a four-month-late 20 min flight, and all went well. Following a 40-min flight on October 14, *L6844* returned to Yeovil, flying again on October 31 after control adjustments. It managed 2½ hrs flying before November 10, when engine bearings seized. By December 11 *L6844* had flown 11 hrs, 4½ hrs in the previous week.

On December 30 1938 four A&AEE pilots each flew the fighter for 40 mins from Boscombe Down, at a flying weight of about 8,830 lb. Take-off was easy but flaps were needed to reduce the landing run and for engine cooling. Stability was good, ailerons light and effective. Minimum speed, flaps and undercarriage down, was 85 mph ASI; best approach speed 100-105 mph. The pilots agreed that the P 9 was a good aeroplane and *L6844* returned to Yeovil, oil cooler troubles needing attention before tail unit resonance tests at the RAE. All well, January 1939 brought the eagerly awaited production order for 200 aircraft and the name Whirlwind. Lord Nuffield's Castle Bromwich factory would build another 800, their cannon to be made at Grantham. Air Ministry confidence

Most unusual in a dark grey finish, the Westland Whirlwind prototype L6844 *in its very early days* (Westland Aircraft).

had increased when John Brown Ltd, ship builders, acquired a controlling interest in Westland during July 1938. Petter transferred their premises to Westland and Associated Electrical Industries invested in the firm. *L6845*, the second Whirlwind, received its first engine in December, the second in January 1939.

Early 1939 saw small modifications to the prototype. Its rudder was enlarged and the rear of the engine nacelles reshaped to reduce buffeting. These modifications delayed *L6845*'s first flight until March 29, by which time the Whirlwind's future had received serious jolts, the first the previous summer. Westland's improved viability coincided with the design of a striking competitor by Supermarine. After their F.18/37 two-motor 12-gun proposal was turned down it was redesigned, keeping the nosewheel undercarriage which had attracted Air Staff interest. The latter saw it as something with which to give other firms 'encouragement' to hasten their work, an aircraft useful if others failed.

Supermarine submitted a brochure describing this revised twin-Merlin six-cannon fighter on August 26 1938. A top speed of 465 mph was claimed by keeping the machine small. Its wing span was only eight per cent greater than the Spitfire's, its wing area 25 per cent more and all-up weight a 12½ per cent increased. Two groups of three 20 mm cannon would be sited in the wing roots for ease of removal and re-arming. The monocoque fuselage and single-spar Alclad flush-riveted, metal-covered wing with fabric control surfaces reflected conventional Supermarine practice. Bristol Taurus engines could alternatively be fitted, wing leading edge root radiators (Westland-style) then being deleted. This version would reach 430 mph at 20,000 ft. Tare weights were 8,769 lb (Merlin version) and 8,088 lb (Taurus), loaded weights being 11,312 lb and 10,536 lb respectively.

Below and opposite *Supermarine Type 327 six-cannon fighter, derivative of the F18/37 Submission. Supermarine first submitted four-cannon tractor and pusher versions of the design, before updating it in 1938 to have six cannon. Details shown: A — six 20 mm Hispano cannon in wing roots, B — fuel tank for 68 gal, each mainplane, C — fuel tank for 42 gal, each mainplane, D — radiator, E — Fowler flap, F — tricycle undercarriage, wheel base 9 ft, G — air intake for radiator, H — air intake to oil cooler on left, to carburettor right, I — wheel doors covering nosewheel after retraction. Close-up of nose, showing nosewheel (32.03 in diameter and mainwheel 36.22 in diameter) and nose shape. Data of August 26 1938 claimed a top speed of 465 mph at 22,000 ft, service ceiling 40,000 ft. All-up weight 11,312 lb with two Merlin 2SM engines. Wing span 40 ft; wing area 304 sq ft, length 33 ft 5 in; height, one blade vertical, 10 ft 3 in. Wing loading 37 lb/sq ft; basic wing section NACA 2200. Total fuel load 220 gal. Alternative armament was twelve .303 in wing guns.*

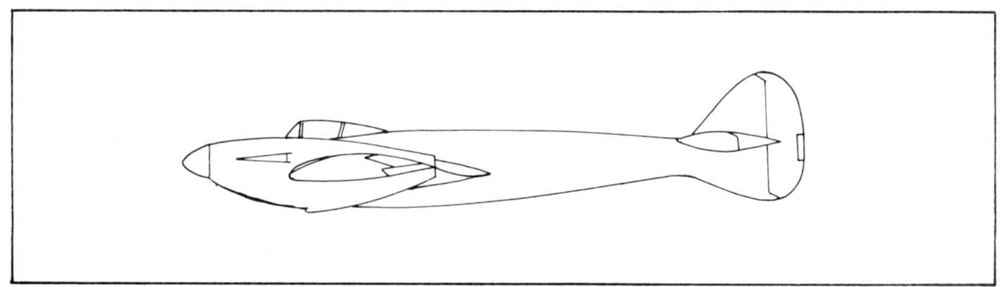

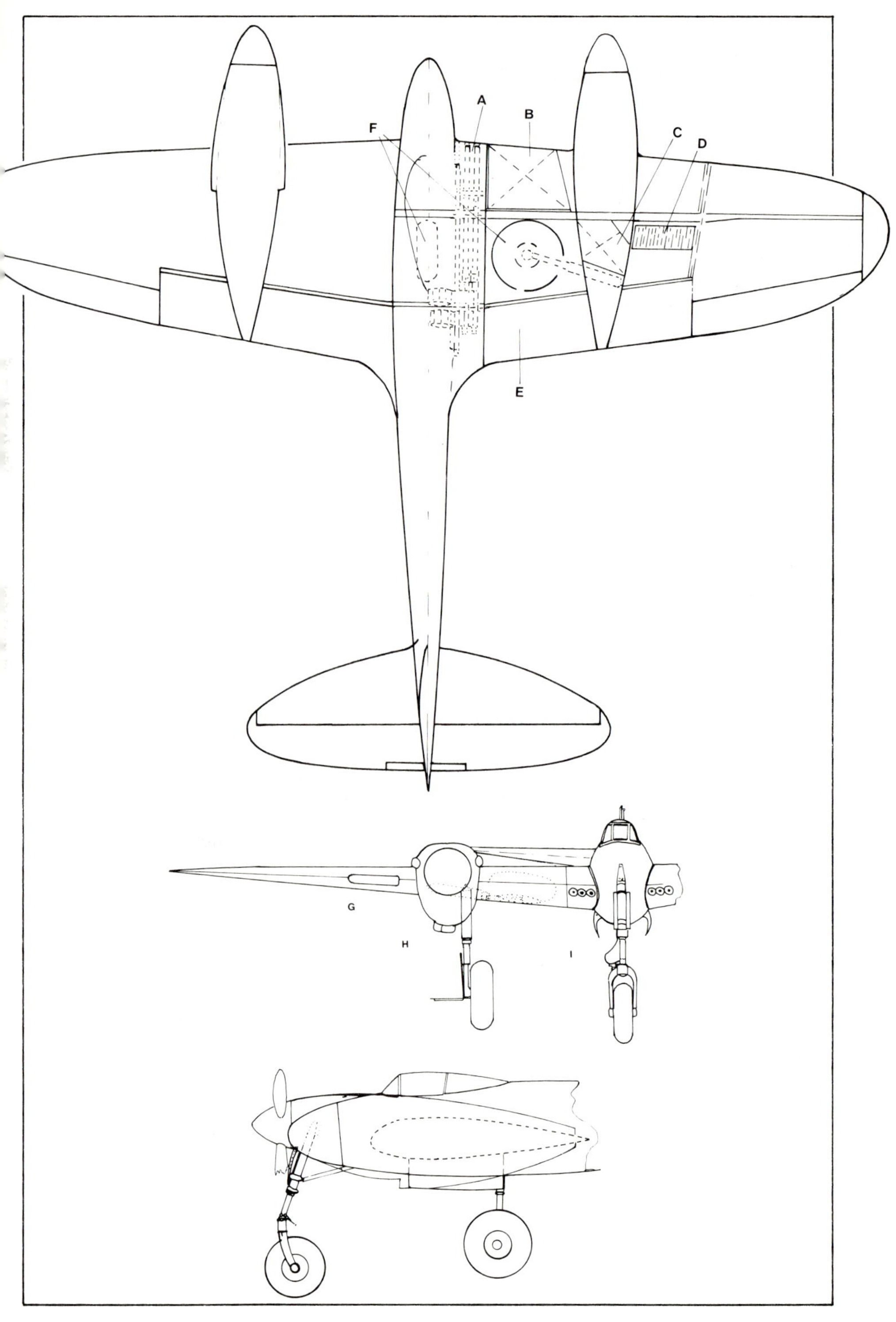

A
B
C
D
F
E
G
H
I

Top *Supermarine's contender for Specification F18/37, replanned as a six-cannon fighter, did not pass beyond the mock-up stage* (Supermarine). **Above** *The Supermarine Type 327 mock-up showing the proposed nosewheel layout and gun ports for wing root cannon* (Supermarine).

Although the aircraft appeared impressive, a detailed tender review was less flattering. Armour had been added, but range reduced by withdrawal of a fuselage tank. A pusher-engined version had originally been tendered, permitting wing root cannon which could have weakened the wing structure. The original design incorporated spoilers for reduction of ground strafing speeds. Additional wing area increased the wetted area, reducing the speed of the new versions. A revised tail unit decreased risks of inadvertent spins, thought considerable with this aircraft which needed slats or careful wing tip design. The Air Staff considered the pilot's view poor and saw no need for another twin-engined type unless it offered advantages — such as space for a gun-loading crew member.

Supermarine's design was discussed on November 24 along with other possible cannon fighters. The Gloster F.9/37 twin-Taurus design, revision of a requirement for a twin-engined fighter with four .303 in guns in a midships

A most attractive aeroplane, the first Gloster F9/37, derived from a turret fighter, would probably have been well worth developing (RAF Museum P6767).

turret, had recently been revised as a single-seater with twin, fixed 20 mm forward-firing cannon. There was a possibility that another three cannon fixed, and slightly angled, could be added. Two prototypes of F.9/37 were expected in the spring of 1939.

What would seriously delay all these designs was the shortage of cannon — and the Grantham factory was working, hopefully, towards building 40 a month.

Chapter 12

Excitement and disappointment

Among F.37/35 designs had been the four-cannon low wing radial-engined Bristol 153 single-seater. Tendered in May 1936, a speed of 357 mph at 12,500 ft was claimed for it, initial climb of 3,580 fpm, service ceiling 33,200 ft. An alternative, Type 153A, had a slim fuselage faired into the wing where its four cannon were grouped. Two under-slung Bristol Aquila radials would give it 370 mph at 15,000 ft. This smaller, twin tailed machine was succeeded by one with buried power plants in thickened wings.

Concern over the pace of cannon fighter development and growing dissatisfaction with the Whirlwind programme were generally known within the industry, but capacity for tackling cannon fighters was limited. Then, with scarcely any warning, an inspired design was born. On October 15 1938 the general reconnaissance/torpedo bomber Beaufort first flew and within days Bristol, (delivering Blenheim Is for conversion into 'auxiliary' fighters) had an idea for a cannon fighter variant, hopefully superior to the Whirlwind and available for production in April 1940. The Air Council hurriedly met on November 29 to discuss this radical notion. A week later the Air Staff decided that Fighter Command must become partially and rapidly equipped with cannon fighters even though no such aircraft were yet available.

Bristol's four-cannon Type 156 had come unexpectedly, yet rapidly securing favour; surprisingly because its progenitor was no glittering performer. To make possible this 16,000 lb fighter, Bristol would fit powerful Hercules engines giving it a top speed of 365-370 mph. Propellers larger than the Beaufort's were necessary which meant a raised thrust line and additional ground clearance. The pilot would have his observer aft to reload the guns.

Bristol, asked for further details, presented them to Air Marshal Sir Wilfrid Freeman on December 23. Beaufort fighters could be available for testing six to eight months following an order. Twin Hercules with high-speed superchargers would confer good performance, their weight being balanced by shortening the Beaufort's nose. Wing strength was more than adequate, and to achieve a fuselage of reduced wetted area retaining many Beaufort components to speed production, its centre portion would be little changed. The slim after section meant a tandem layout. Increased fuselage base width would permit two cannon each side of novel entry/exit doors between. More careful performance investigation suggested a top speed of 361 mph at 15,000 ft. Cruising at 314 mph at 15,000 ft, duration would be 2.42 hrs, equating a range of 760 miles — or 1,345 miles at 150 mph, duration then becoming 8.97 hrs.

Further Air Council discussion on January 2 1939 echoed Bristol's enthusiasm. Already the firm had embarked upon a serious design study, L.G. Frise, Chief Designer, pointing out that Bristol had cannon experience and were working on a twin-cannon Blenheim against a foreign order. Delay would mean that it would be September before the first aircraft was delivered, although production of a '370 mph fighter' could still commence in early 1940. That would be eased because of the many components common to the fighter and the Beaufort. Bristol's claims were readily accepted because the company was considered dependable — and because of the increasingly urgent need for a cannon fighter.

Bristol suggested that 100 fighters could be built by December 31 1939, 235 by a year later. Completion of 350 RAF Beauforts by February 1940 would allow a full switch to fighters following the phase-out of Blenheim production late in 1939. Current plans called for Bristol to start producing Stirling bombers in February 1940, and the suggestion now was for the company to build the fighter instead. On February 7 1939, meeting at Filton, the Air Council decided to do that, and go ahead with a £3.3 m order for 300 cannon fighters. Included would be two prototypes, hastened by the use of components from the 71st and 72nd Beauforts.

At January's design conference it had been agreed not to put out to tender the aircraft's specification. To speed production, early aircraft, perhaps 55, would have Hercules II engines. The Hercules VISM with higher altitude rating should then be available. Performance limitations were accepted — with some concern — when the requisition order for 300 fighters was formally signed on February 24 1939. Development of a specialised night fighter also seemed unnecessary now, for this large aircraft could accommodate much equipment and fly lengthy patrols. Exceptionally roomy, it might even carry an additional crew member.

Knowing that Petter and Westland were concerned at these developments, the Whirlwind designer was invited to view the Bristol mock-up. 'Far too large,' he commented, 'especially the fuselage.' Frise replied that in two year's time such size would be necessary — and he was right. More important, the Bristol machine had development potential denied to the too-small Whirlwind.

Bristol's engine production was arousing increasing concern and when on March 11 1939 a statement of operational requirements for the Beaufort fighter reached Bristol it categorically stated that the aircraft must be fitted with Hercules VI engines, that its top speed must be 360 + mph at 15,000 ft and that four cannon were needed. Conversion of two Beauforts had started and the prototype was to fly within nine months. By the end of December 1939 hand-made pre-production aircraft would also be flying.

Official examination of the Bristol mock-up on April 17 1939 led to detailed provisions, embodied in Specification F.17/39 of April 27 1939, for an aircraft commonly called, since December 1938, the 'Beau Fighter'. The specification sent to Bristol on May 17 referred to it as 'an interim replacement for F.37/35 before full replacement comes', showing an early readiness to end the Whirlwind.

Below 15,000 ft the Beau Fighter's speed was ordered to be 'not less than 330 mph at 5,000 ft'. It must reach 20,000 ft in 10 mins, so that some lowering in performance was accepted as inevitable. Self-sealing fuel tanks were needed and protective armour. The Air Ministry was determined that Hercules VIs would soon be fitted but the Hercules was in trouble. By the end of April 1939 Bristol reluctantly informed the Ministry that many early Beau Fighters must have Hercules IIIs. This meant that the original speed demanded could certainly not be met, forecasts now showing 336 mph at 15,000 ft and 318 mph at 5,000 ft.

Now Gloster's small F.9/37 cannon fighter emerged. It stemmed from a twin-engined turret fighter (F.5/33) updated to F.34/35 by combining a four-gun dorsal turret with two nose cannon. A third revision resulted in F.9/37, a two-motor, all-metal, stressed-skin single-seater day and night fighter with twin nose cannon. Construction began in February 1938. On April 3 1939 the first example, *L7999*, flew. It exhibited good handling qualities and a top speed, when not quite fully loaded, of 360 mph at 15,000 ft. On July 27 1939, *L7999* flew to Martlesham. A week later it was damaged in a crash landing, and returned to Gloster on August 2. Bristol Taurus air-cooled engines powered the fighter, notoriously troublesome power plants, development of which was directed towards making them successful at low altitude for the Beaufort torpedo bomber. During repairs *L7999*'s 1,050 hp two-row 14-cylinder Taurus T-S(a) engines were replaced by the only developed version, the 900 hp and less supercharged T-S(a)3 with which many problems were to arise. Less power reduced the loaded aircraft's speed to 332 mph at 19,200 ft. Taking 6 mins to reach 14,000 ft, it remained pleasant to fly.

For the second prototype Gloster turned to the only alternative small engine, the 880 hp Rolls-Royce Peregrine whose future was bleak. Eventually, *L8002* flew on February 22 1940. Heavier than the forerunner (12,108 lb compared with 11,653 lb) it attained 330 mph at 15,000 ft when fully loaded. The F.9/37 was reckoned too lightly armed, too small for development, although it re-entered the fighter stakes later in 1940.

Meanwhile the Beaufighter's performance had been further diluted because the first 30 were to have Hercules II engines, poorly rated above 10,000 ft. An Air Staff meeting on June 22 1939 received this disturbing news and replied with a revised specification requiring a minimum top speed of 336 mph at 15,000 ft and a climb rate of 2,080 fpm in Hercules III aircraft. Bristol replied stating that there was 'no real cause for concern', 361 mph at 15,000 ft would be attained with the Hercules VI and a climb rate of 3,680 fpm. Relying upon 100 octane instead of 87 octane fuel, and a higher supercharger gear ratio — an embodiment difficulty which Bristol overlooked — the company had underestimated the time scale and problems the Hercules VI would encounter. The Air Ministry realised that the Mk VI would need lengthy development and became resigned to having many Hercules III aircraft. Despite deep concern, the idea of cancellation was never mooted, for a cannon fighter was desperately needed. Performance-wise it would become inferior to the Hurricane and production delays were thought certain. Nevertheless, because of the seriousness of the international situation, the order for 300 Beaufighters was confirmed on June 26 1939, likewise the aircraft's name which had already slipped into usage.

Neither of the cannon fighters proved much superior to Hurricane *L1750* which, on May 24 1939, made its first flight and with a bulky cannon beneath each mainplane. Hawker had been nudged into the cannon age by an Air Ministry belief that ingenious Supermarine would somehow produce a cannon Spitfire. According to observers of those days, the Ministry was determined that its prime fighter supplier would not be left out. Hawker went along with this, but the guns' slow rate of fire and limited ammunition reduced the aircraft's

Cannon Hurricane prototype, L1750, *carrying a 20 mm Hispano cannon externally beneath each wing for trials (BAe).*

popularity among fighter pilots increasingly accustomed to eight guns. *L1750* went to Martlesham for comparative tests against an eight-gun, wooden-propeller Hurricane in June 1939. They showed *L1750* to handle as well as any at 6,169 lb flying weight (4,847 lb tare). It reached 302 mph at 16,800 ft, (about 8 mph less than a normal Hurricane), took 7.6 mins to reach 15,000 ft and 18.1 mins to attain 26,000 ft. The Air Ministry promptly approached Hawker to fit four cannon but not until June 1940 was this initiated.

By summer 1939 Bristol were complaining about the lack of precise instructions concerning the Beaufighter's internal layout. This had been intentional so that Bristol were not inhibited. Greater concern came with increased awareness that the torpedo bomber's fuselage was far different from the fighter's. Indeed, it needed entirely redesigning, something not initially intended, and which would promote more delay. Bristol did well to fly *R2052*, the prototype, on July 20 1939.

Following the cannon Hurricane's trials, Hawker, in July 1939, received an order for two sets of four cannon F.18/37 wings, although cannon production was earmarked for 300 Beaufighters and 400 Whirlwinds. Wing modifications would be far from simple, but the Air Staff decided that one of each F.18/37 prototypes must mount cannon. The CAS would not, however, agree to production F.18/37s having cannon until the value and reliability of the weapons was proven beyond doubt — and that did not come about until late 1940.

Westland were refining their Whirlwind against the production specification F.37/35/PI approved on June 5 and sent to the company on June 17 1939. The second prototype was having modifications to the oil-cooling system, armour installed, a stiffened fin and a rudder change. Comparative trials with the first prototype followed for, trying a French idea, the latter had engines/propellers rotating in opposite directions whereas those of the second machine rotated one way. Tests showed the additional complication of handed engines was not worthwhile. The Air Ministry, by refusing to order Peregrines until these tests were undertaken, had further delayed production, now set at eight in February 1940, 12 in March and 22 in April, increasing to a peak of 48 in December 1940.

Uncertainty concerning the Whirlwind was reflected at the ACAST's Liaison Meeting of March 23 1939, when production at Castle Bromwich was abandoned in favour of Spitfires and another 200 Whirlwinds from Yeovil. Some 1,850 fixed-gun fighters were needed of which 300, or possibly 600, would be Bristol twins. Another 800 fighters would come from Gloster or Westland. To build 40 Whirlwinds a month by June 1941 meant additional capacity at Yeovil.

The Air Ministry's demands for many modifications worried Westland. Petter protested, to little avail, and realising their fighter could not enter service until 1940 they up-dated it as much as possible, which meant repeated delays. The company was told its fighter needed tropical equipment, which was not in the specification. To hasten production the Ministry agreed that the first 24 aircraft could be completed for Home Service only and on July 1 1939 production commenced.

The first nine Whirlwinds, non-standard, were to have been set aside for experimental purposes. When they became available, though, operational needs were such that they were delivered to the first squadron. The 10th to 25th aircraft, although better equipped, were still not to full operational standard. One modification which constantly brought problems was a small door in the rear fuselage, which upset skinning. Only tropicalised aircraft needed it . . . and the Whirlwind was destined never to serve overseas.

Throughout 1939 the Air Ministry openly maintained that Westland was devoting too much time to the Lysander. Privately it was convinced that the company was too small to handle a major fighter programme. Rolls-Royce, knowing their Peregrine had less development potential than the Merlin, was not developing it at the rate Westland hoped for. Fuel supply was being temperamental, the carburettor priming pump circuit, with non-return valve in parallel to engine fuel pumps, was causing trouble. Vapour lock, due to excessive temperature, was another problem.

How little the Air Staff favoured the Whirlwind was obvious on August 4 1939. Production outlines for F.17/39 went to Bristol, carrying the pre-amble 'the Air Staff requires a fixed-gun fighter, with a number of 20 mm cannon in service as quickly as possible to replace the F.37/35'. It continued 'a proposal based on the Beaufort has been accepted. It should have Hercules HE VISM engines' (the Air Staff refused to refer to the Hercules III), 'and a speed at 15,000 ft as high as possible and not less than 350 mph. Service ceiling with full load must not be less than 30,000 ft, and the aircraft must attain 20,000 ft in 10 mins. Its four cannon must be able to fire forward at 300 mph at 15,000 ft, and be able to be reloadable in the air. A mechanical method of conveying the 16×60 round ammunition drums to the guns is required.' By this time only the outer wings, tail unit and undercarriage were recognisable as parts of the Beaufort, so radical had the alterations become. All too obvious, when September 1939 arrived, was Bristol's broken promise of delivery. The second Beaufighter, *R2053*, did not fly until November 22 1939.

On September 1 1939 Whirlwind *L6844* proceeded to the A&AEE for gun firing trials and attempts to simplify its complicated maintenance schedule. The Air Ministry discovered that many requested modifications had not been subjected to trials so the prototype was returned to Yeovil on September 16 for their incorporation and testing, flying resuming on November 27 1939. Jigs were in place when war started, first mainplanes were completed in October and a fuselage and wing during December. At an ACAST meeting of November 1 1939 an unpursued suggestion was for Whirlwinds to replace Spitfires as photographic reconnaissance aircraft. With additional internal fuel tanks, and guns removed, they would have become available by July 1940.

The first two production Whirlwinds were assembled in February 1940, their four 20 mm Hispano Suiza Mk 1 cannon each having 60 rounds. Twelve Browning guns remained a possibility, encouraging ideas of mixed and increased armament. The pilot's view from the advanced-style bubble canopy was good. He was protected ahead by a 9 mm steel armour plate bulkhead across the fuselage and shielding the cannon drums, a bullet-proof windscreen and a thick alloy deflector panel fitted on the fuselage top. Rolls-Royce Peregrine engines, which Petter maintained should have been produced handed, were giving 805 hp at 15,000 ft.

Certainly the Whirlwind was novel with radiators cunningly slipped, reducing drag, into the wing leading edge. The undercarriage was fully retractable and the rear of each nacelle neatly slipped forward as well as downward to serve as a flap, reducing wing area and therefore acting as an efficient high lift device. Fowler flaps were fitted, in conjuction with Handley-Page slots to aid slow flying and stall characteristics. Wing loading was 40 lb/sq ft. These features and the heaviest gun armament of any production British fighter made the Whirlwind a pace-maker. But its future remained increasingly insecure because of Peregrine engine troubles. Questions were asked about whether the aircraft embodied too many innovations, many of which were commendable and worked quite well. Had the Whirlwind been larger and stronger and laid out for two Merlins, by how much would it have differed from the Mosquito?

Although increasingly winning favour, the Beaufighter *R2052* took longer to prepare for official trials than expected, adjustments following each flight. Eleven flying hours were erratically accomplished, some in quick succession, some with weeks between. Not until January 1940, six months after the prototype first flew, did it venture to Boscombe Down for trials. Jigging, tooling and design complications were holding back production and engine problems further delayed the Beaufighter.

Shortly after the war began agreement was reached on a large scale order for Merlin-engined Beaufighters as an insurance against major Hercules problems. Wright Cyclone and Griffon engines were alternatives to the Merlin engines of the Beaufighter II, delivery of which was scheduled for spring 1941 when a breakdown in Hercules production seemed likely in August 1940. Three airframes were ordered to be 'sacrificed' as Merlin-engined prototypes.

Westland were only too well aware of their fighter's position so prepared the Whirlwind Mk II, presenting plans to the Air Ministry on February 12 1940. Westland claimed that Rolls-Royce were improving the Peregrine to accept + 12 lb boost at 20,000 ft. This 15 per cent increase would add 40 mph to the Whirlwind's top speed. Engine revs in fast dives being high, a propeller pitch increase from 20° to 35° needed development. New radiators would be of the Morris Film Type. Rounds per gun would be raised to 120 and Petter's other suggestions included supplementing the Mk I's two 67-gal tanks with a 25 gal fuel tank aft of the cannon bay and a 35-gal rear fuselage tank.

Within a week the Air Ministry reminded Westland of the promise — Whirlwinds in production nine months after the January 1939 order. It was to take four years, four months to reach delivery state. The Mk II, already outdated, lay too far ahead. No 100-octane Peregrine existed and + 12 lb boost could only be achieved in emergencies. In March 1940 a pair of Morris fuel and tube radiators were applied to a Whirlwind adding 5 mph to its speed. Another improvement officially suggested, was continuous cannon feed, requiring lower-sited guns and 120-round containers gravity feeding into the breeches. This provided space for an additional 27-gal fuel tank, a 20 per cent increase, extending cruise range from 720 to 900 miles.

Westland did not easily give in, soon presenting performance data for a 100-octane Peregrine Whirlwind carrying 480 rounds of ammunition, comparing it with the normal Mk I:

	Mk I 4 × 20 mm cannon, 240 rounds, 87-octane fuel	**Mk II** 4 × 20 mm cannon, 480 rounds, 100-octane fuel
Top speeds (TAS)		
5,000 ft	342 mph	354 mph
20,000 ft	369 mph	422 mph
25,000 ft	371 mph	411 mph
Time to:		
15,000 ft	7.1 mins	5.5 mins
20,000 ft	10 mins	7.4 mins
25,000 ft	14.5 mins	9.6 mins

Engine power would increase from 1,770 hp at 15,000 ft to 2,030 hp at 20,000 ft, loaded weight rise from 10,020 lb to 10,465 lb. If none of this was acceptable Westland hoped to at least build a new nose section interchangeable with the usual one. A projection of May 1940 was for one carrying four cannon, three machine-guns and a 30-gal fuel tank.

Official interest was more concerned with the Beaufighter; 'concerned' is the operative word for when the first A&AEE performance reports had reached London on February 6 1940 they were, indeed, disturbing. *R2052* had a top speed, when fitted with Hercules 1-SM engines similar to the forthcoming Mk III, of 335 mph at 16,800 ft. *R2053* with Hercules 1-M, similar to the Mk II to be fitted to initial production aircraft, returned a mere 310 mph at 4,000 ft. It was not long before *R2054* was tested with Hercules IIIs giving 1,270 bhp at 15,000 ft, with top speed an alarming 309 mph at 15,000 ft. Examination of this machine on February 19 1940 shocked the authorities because of the poor finish. When its troublesome, retractable tailwheel was locked down and a

The third prototype Beaufighter I (Hercules III) in early wartime fighter finish, during flight trials at Filton in 1940. It joined 604 Squadron on January 19 1941 and was destroyed on crashing during approach to Middle Wallop on February 6 1941 (RAF Museum P014327).

camera gun blister was fitted above its canopy the top speed fell to 300 mph at 14,060 ft. *R2054* had a revised engine nacelle shape with wheels further protruding after retraction and it needed finer cockpit framing for night flying purposes. Six early aircraft would have Hercules IIs and the third and sixth prototypes would be used for radar, radio, cannon firing, cockpit heating trials, etc.

New, influential support came to the Beaufighter's rescue on March 4 1940 when Coastal Command first expressed interest in a three-seat low-level strike and shipping escort version, with six additional machine-guns; four in an aft turret and two ventral for attacking shipping. Requirements of such a Beaufighter reached Bristol on April 23, the suggestion being that the 50th aircraft could carry the fittings and that, after two fighter squadrons had equipped, a Coastal Command squadron would receive modified Beaufighters.

During April 1940 the first four production Whirlwinds were assembled. Plans were revived on April 17 for the lightened photo-reconnaissance version operating at 30,000 ft. A fixed-undercarriage variant for long range maritime patrol was also considered. By mid-May calculations revealed that a PR Whirlwind might be able to achieve 373 mph at 25,000 ft and manage a 1,000-mile journey at 32,000 ft, although superior aircraft such as the Mosquito and Spitfire PR III seemed preferable. Future prospects for the Whirlwind, already bleak, received a resounding blow on May 8. The company was told that, because Rolls-Royce were busy on the Merlin and Vulture, Peregrine production would only match the needs of 114 Whirlwinds, and would cease in December 1940. Further Peregrines could only be produced by a 2:1 reduction in Merlin output, or by postponing Griffon production set to commence in July 1941. Rolls-Royce claimed that a Peregrine boosting the Whirlwind's performance to that of the projected Mk II 'would require 400 octane, not 100 octane' fuel! Agreement was made to provide for brief boosting, and to review possible use of 100 octane fuel.

Traumatic events across the Channel in May 1940 resulted in the establishment of the Ministry of Aircraft Production and led to Lord Beaverbrook being put in a position of great power over aircraft design,

L6844 earned its keep in later life by undertaking a variety of mixed gun trials, and spent the last six months of its active life at AGME Duxford. Compared with its original style, it can be seen to have different exhaust stacks, radio masts, fin bullet, fixed slats and no mudguards, all of which could have an appreciable effect upon any fighter (Westland Aircraft).

development and production. Beaverbrook selected an emergency aircraft production programme listing three bomber and two fighter types, after which the Beaufighter had first priority. Immediately, Sir Wilfrid Freeman wrote to Lord Hives at Rolls-Royce instructing the company to build 100 Merlin XX power eggs for delivery in December 1940 to allow Beaufighter IIf production to commence.

Only minor improvements to the Whirlwind would now be permitted. Westland received a contract in May to produce the proposed low cannon mounting, in a nose built on a trolley and were told on May 17 that a continuous cannon feed system should be fitted as soon as possible. At the end of May the first production Whirlwind flew. In June *L6844* had a cannon fitted experimentally with continuous feed, but fitting three Brownings fell from favour. Maker's performance forecasts were showing the production aircraft's speed at 17,000 ft close to the Spitfire's, whose superior supercharger showed advantages when higher. Much faster than the Beaufighter, a 10 mph advantage might be retained over the Hercules VI type. A&AEE trials with *L6845* indicated that Westland's claims were inflated.

Loaded to 10,072 lb, 354 mph at 15,800 ft was reached. It took 6.3 mins to attain 16,500 ft, 8.2 mins to 20,000 ft and 24.6 mins to 30,000 ft. Take-off to clear 50 ft took 710 yds; landing 635 yds. *L6845* joined 25 Squadron on May 30 1940, for intensive trials at Debden and North Weald, *P6966* and *P6967* following in mid-June. Teething troubles and a decision to replace Blenheim If fighters by Beaufighters brought a decision to re-form 263 Squadron as a Whirlwind trials formation based in Scotland. Although aware of Beaufighter problems, the Air Staff confidently decided to equip a squadron with 16 + 2, including two aircraft intended for ammunition feed clearance. Eager for battle-ready Beaufighters, they cut the 100-hr intensive flying scheme and paid a high price the next winter for their impatience.

On July 26 1940 the Beaufighter was declared fit for operational service with 25 Squadron. Complex equipment, new engines and troublesome cannon all needed frequent removal. More disturbing was the failure to devise effective ammunition feed for cannon which troubled all fighters. Weak gun mountings unable to withstand recoil brought poor accuracy. Beaufighter delivery was faster than the Whirlwind's, *R2054* and *R2055* in June, five in July, 23 in August, 15 in September and 65 more in 1940.

Lord Beaverbrook, on June 28, after the Whirlwind's cancellation, asked Air Chief Marshal Sir Hugh Dowding, Commander-in-Chief of Fighter Command, for his opinion of the aircraft. His three examples were all grounded, but 'In their favour', he wrote, 'I can say that pilots like them, but this may be due to their relief at getting away from the Blenheim. They may be worth their weight in gold in the near future as being the only fighter aircraft capable of attacking tanks from the air. Against them there is their very high approach and landing speed. Pilots tell me they have to bring them in at 110 mph and even then the controls are sloppy. This means they can probably never be used at night.' This is one reason why 25 Squadron lost them. 'If the Lysander is any criterion', wrote Dowding, 'we can expect an infinity of troubles with cowlings, fittings etc, until the firm has learnt wisdom from experience. Further, I think it is a very extravagant design. By this I mean it takes two engines to lift four cannon whereas the new Typhoon should be able to lift six with one engine. We may, however, be glad to get as many as we can in future, both as anti-tank weapons and to attack bombers as they become less vulnerable to rifle calibre fire. We shall be glad enough to replace them when the Typhoon comes into heavy production.'

Production delays at Yeovil caused Eric Mensforth to refer to problems with fuel tanks and engines unable to hold boost pressure — and one which ran away to 4,000 revs. Production was now set at four Whirlwinds in July, six in August, eight in September, ten in October and a dozen monthly thereafter. Tropical requirements were cancelled and during August 1940 pilots of 263 Squadron at Grangemouth, equipped with Whirlwinds, encountered alarming experiences while learning to master fast moments and realised that this was not the ideal cannon fighter. That would evolve from adaptations of other aircraft.

Chapter 13

Adaptations and setbacks

Spring 1939 found the Hawker F.18/37 way off, doubts about the Whirlwind and the wisdom of making a torpedo bomber into a fighter becoming of increasing worry. The Air Staff and Ministry thus had little choice but to cajole their respective makers into producing cannon-armed Hurricanes and Spitfires. They opted for the thick-winged Hurricane and informed Supermarine that cannon must, somehow, be wedged into Spitfire wings. Slinging cannon beneath the Hurricane's wing and, hence, reducing its performance was an interim answer. Supermarine, ever ingenious, fitted into each mainplane of Spitfire *L1007* a cannon lying on its side. Each gun's 60 round magazine was accommodated beneath a streamlined blister. *L1007* flew to Martlesham in late June 1939 for firing trials over Orfordness. Feed problems were repeatedly encountered due to spent cartridges blocking the gun breeches, a problem partly due to the angled gun setting and partly due to weak mountings. Clearly, the Spitfire could be cannon-armed, but only with some difficulty. There was talk of building 250 cannon monthly, even 750, making non-cannon armed fighters doomed. Some, though, argued that dense concentrations of fire remained preferable to a few heavy shots from a cannon.

Far from service-ready, twin-cannon *L1007* reached AFDE Northolt on November 9 1939 and then, operating in conjunction with 65 Squadron, was despatched for trials off the north-east coast in December 1939 during which she test-fired explosive ammunition. Favouring cannon, the Air Staff on December 6 1939 decided to arm the first 200 Tornadoes with cannon until it was realised that sufficient guns would be available only for the first 90. *L1007* was at Drem when, on January 13 1940, a scramble was ordered, a Heinkel He 111 being off the Scottish coast. Spitfires of 602 Squadron were attacking it when *L1007* in the hands of Flying Officer G. Proudman joined in; later it was considered that his fire contributed significantly to the Heinkel's demise. Output of the guns increased appreciably when BSA Sparkbrook began, on April 12 1940, to deliver them for Whirlwinds and Beaufighters.

On March 6 the Air Staff optimistically decided to arm the Tornado with six cannon, intending this to be the normal cannon fighter armament. An investigation by Hawker revealed that the additional 220 lb involved would require wing-torsion examination (or) re-evaluation — and just as tooling for the four-cannon wing was commencing. Twelve .303 in Brownings comprised 850 lb of the aircraft's load, while six cannon weighed 1,300 lb. To reduce this weight, the Air Staff suggested 100 rpg instead of 135 rpg, two sets of experimental six-cannon wings being ordered on high priority. Whatever the result of their experiment, it could not alter the slow rate of cannon delivery. In Churchillian style, Lord Beaverbrook added his support to the six-cannon lobby. Best to hit the enemy as hard as possible in one go! Cannon were Beaverbrook's preference and, squashing pleas for caution, he secured agreement for 30 Spitfire Is to be fitted with cannon in the *L1007* manner. They were to be embodied during production, final touches coming at 6 MU Brize Norton — and delivery would take place before full A&AEE trials — which was unwise.

At Fowlmere late in the evening of July 1 1940, Squadron Leader P.C. Pinkham, commanding 19 Squadron, the premier Spitfire squadron, revealed it would be equipping with cannon Spitfires, three of which had arrived at Duxford. Unreliable feed and poor cartridge-ejection remained, contrasting with the increased rate of fire of the Browning II (Star) currently entering service. Six seconds of cannon fire seemed a short time to the pilots who foresaw only one pass as practicable. They would be depending upon the cannon's high muzzle velocity decreasing deflection and increasing destructive power.

Soon confirmed was an improvement in firing accuracy. On July 3 No 19 Squadron returned to Duxford, the weeks ahead leading to a new slant to fighters of the future. On July 4 a different manoeuvre was tried. Sections flew in echelon 2,000 ft above and abeam of the enemy then, holding aim, each aircraft dived on its target before climbing away to hold before a second strike. But the enemy habitually flew just below cloud level, decreasing the value of such tactics. Next day's firing practice from Sutton Bridge brought plentiful stoppages resulting in the fitting of a rubber pad to the spent-round deflector plate, preventing it from rebounding to be trapped in the forward-moving breech block. Special gun sights were also fitted.

Cannon Spitfires commenced operations on July 9, flying convoy patrols from Coltishall and Martlesham. No engagements took place in July. The nearest to success the squadron came was on August 2, when Blue Section led by Flight Lieutenant W.G. Clouston intercepted a He 111 of KG 55. Only the leader fired, as the other two had stoppages, and he had an eight-gun Spitfire. Enthusiasm waned fast. When, on August 11, a two-cannon four-machine-gun Spitfire joined the squadron there was disappointment that it did not have eight machine-guns.

Shipping patrols continued and late in the afternoon of August 16 'A' Flight was ordered to investigate an enemy formation off Harwich comprising 150 bombers, 40 Bf 110s of ZG 26 stepped up to the rear and some Bf 109s providing top cover. Outnumbered, the seven Spitfires went for the Bf 110s. In a fast fight, three were listed as destroyed and one seriously damaged. Two were claimed by Flight Sergeant G.C. Unwin flying *R6776*, and six Spitfires had stoppages. Unwin, almost certainly, was the first RAF pilot to alone shoot down a raider with cannon.

Two cannon-armed Spitfires destroyed a Do 17Z of 7/KG 2 over Essex on August 19 and in mid-afternoon on August 24 No 19 Squadron engaged Messerschmitts escorting a raid on North Weald. 'A' Flight claimed three Bf 110s whilst 'B' Flight, facing Bf 109s, found that the guns of only two Spitfires would fire. An attack on Debden materialised on August 31, Do 17Zs escorted by Bf 110s heading for the famous airfield. Cannon Spitfires were ordered to engage the raiders, Sergeant D. Cox in *R6924* claiming a Bf 110 and Flight Lieutenant Clouston another, which he shared with Pilot Officer E. Burgoyne.

Flying Officer J. B. Coward in *X4231* was shot down and Flying Officer F. Brinsden's *R6958* fell to a Bf 109 near North Weald. A further loss came when Flying Officer R.A.C. Aeberhardt was killed making a flapless landing in *R6912*. Litle wonder that 19 Squadron withdrew to Digby on September 2 to equip with conventional Spitfires. For the remaining 23 cannon Mk 1s, though, it was far from the end; they would play a crucial role in Spitfire development.

Interest in mixed cannon/machine-gun armament again arose. The Spitfire easily accommodated this, having a superior gun layout to the Hurricane. On August 28 1940 came the decision that Spitfire IIIs would feature a 'Universal Wing'. Two cannon and four machine-guns would be carried with a possibility of more machine-guns if cannon production slumped. Beaverbrook leaned hard on Supermarine to fit effective cannon into Spitfires and a third cannon factory commenced production on August 27, an event celebrated by talk of fitting four cannon into Spitfire IIs and IIIs following the mixed complement.

A further use for cannon was suggested in June 1940 as the invasion of Britain seemed likely. A committee met on June 26 to discuss the usefulness of cannon against tanks. Bristol's mock-up of twin cannon beneath a Blenheim IV was recalled, but finally it was decided to issue to each Lysander squadron a few 20 mm cannon to be lashed to bomb stubs for ground attack use. They would be withdrawn as soon as practicable. An idea bringing home the nation's peril (or to add a note of humour?) was a suggestion for 37 mm cannon to be installed in antique Vickers Virginias which would tour our coastal resorts as necessary, fire upon invaders and probably frighten friend and foe by displaying Harrier-like ability to fly backwards on windy days. One serious soul at the meeting pointed to a shortage of 37 mm COW guns, claiming that stocks were too low to warrant the necessary trials and development this fascinating project demanded.

What was true of the 'Ginny' was most untrue of the Hurricane. Hawker received a go-ahead early in June to install four 20 mm cannon within battle-damaged wings. Layout decided, P.1002 Hurricane *V7360* had the wings fitted and flew to the A&AEE on August 3 for firing trials, which were completed on August 15. It joined 151 Squadron at North Weald on August 19 for operational trials, the general impression being that speed was slightly sacrificed for a hefty weapons punch. On August 22 it took off at 08.20 hr for its first of ten patrols. Flight Lieutenant Smith during an early sortie on August 31, claimed a probable Bf 109 and next day *V7360* was switched to 56 Squadron which maintained the aircraft while 151 Squadron was at Stapleford Tawney. Between September 3 and 13 it again served in 151 Squadron but flew no operations. *V7360*, unrepresentative of production four-cannon Hurricanes, prompted September 2's demand for 30 such conversions by mid-October, no more feasible than the call of August 29 that 30 become available before October 12! Instead, and using damaged wings, the Hurricane's cannon-feed system was redesigned. By September 15, eight Hurricane IIs were each being fitted with four cannon. Long protruding gun barrels called for aerodynamic refinement, after which *V7360* proceeded to the A&AEE on February 22 1941. Two months later it moved to Farnborough from where, on October 3 1941, it crashed. Its 1940 success had caused the Air Staff to confirm, on October 6 1940, that they wanted all production Tornadoes and Typhoons to have cannon.

Blighted by engine troubles, cannon-feed problems and dented nose cones resulting from cannon blast, No 263 Squadron's Whirlwinds were by then at

Drem. Larger, slower than expected and vulnerable, the Beaufighter was also in trouble although its entry to service came fortuitously just as the Luftwaffe opened its main night-bombing offensive. The Beaufighter's capaciousness ideally suited it for bulky AI radar, the cannon loader being well sited to act as radar operator/observer. Handled by a well-matched crew, the Beaufighter proved itself an excellent night fighter in a role for which it had not been conceived. That huge fuselage must surely have become the envy of Petter. The Air Staff already wanted a faster night fighter, but speed was not as yet all that important in a night fighter.

In May Gloster had their F.9/37 flying again, in an attempt to resurrect official interest. On May 29 Mr W.G. Carter, Gloster's designer, wrote to the MAP suggesting a revised Taurus F.9/37 as ready for immediate production. The MAP considered that idea as remote, and that Gloster should concentrate upon jet-propelled aircraft. Nevertheless, they suggested a Merlin XX version might be of use. Carter's team worked overtime and, two days later, offered a 14,500 lb, 1,000 mile-range cannon fighter. He heard nothing for a month, after which it transpired that a filing error had interrupted correspondence. When the answer arrived it contained news that because of good reports of the second F.9/37 the MAP was interested, although wondered if this would be the best application of Merlin XXs. By late July the Merlin XX Gloster F.9/37 Reaper cannon fighter had become an important item, especially since it coincided with designs for Specification F.18/40 calling for a specialised cannon-armed night fighter to replace the Beaufighter. 'F.9/37 would be a very good answer to F.18/40', claimed the MAP, especially if fitted with five cannon and air-cooled engines. The DTD, however, felt that a hybridised, spoilt F.9/37 would be the only outcome.

Replacement for the Blenheim If and Beaufighter was considered on July 16 1940. Slow, and with a poor view, the Blenheim was outdated. Heavy cockpit framing and huge engine cowlings were undesirable features of the Beaufighter. Therefore a fast, two-seater night fighter was formulated. Discussion ranged around the value of a turret which Wing Commander King claimed made attacks from below possible. Complicated to design, this would make the aircraft slower, he was told, and the meeting settled for a fixed-gun machine. Six cannon seemed too much, and the value of mixed armament was questionable. Six .303 in guns were suggested, before four 20 mm cannon were decided upon. The ACAST pointed out that the Typhoon would have six cannon. The meeting then decided that F.18/40 should do likewise. Issued on August 23, the specification called for rapid production, wood or composite material being acceptable. To achieve 400 mph at 20,000 ft, a 3-hr patrol at 15,000 ft and a ceiling of 35,000 ft, two-speed engine blowers were essential. Take-off run needed to be 500 yds, landing run 650 yds. Armament would be six cannon, or four cannon and six .303 in guns.

September brought novel F.18/40 submissions. Fairey devised a single-engined turret fighter, AI-equipped, a strengthened N.5/40 Firefly derivative having two wing cannon. Boulton Paul offered a Sabre/Griffon scaled-up Defiant. Using a Sabre NS3, they forecast a speed of 389 mph at 21,000 ft or 367 mph if a turret was fitted. Instal a high-altitude NS6SM and speeds at 34,000 ft would be 415 mph and 392 mph respectively. No single-engined fighter could meet the need. Boulton Paul tried again, submitting ideas for a radical twin-boom twin-Sabre layout. It was 'too futuristic'. The MAP decided, on

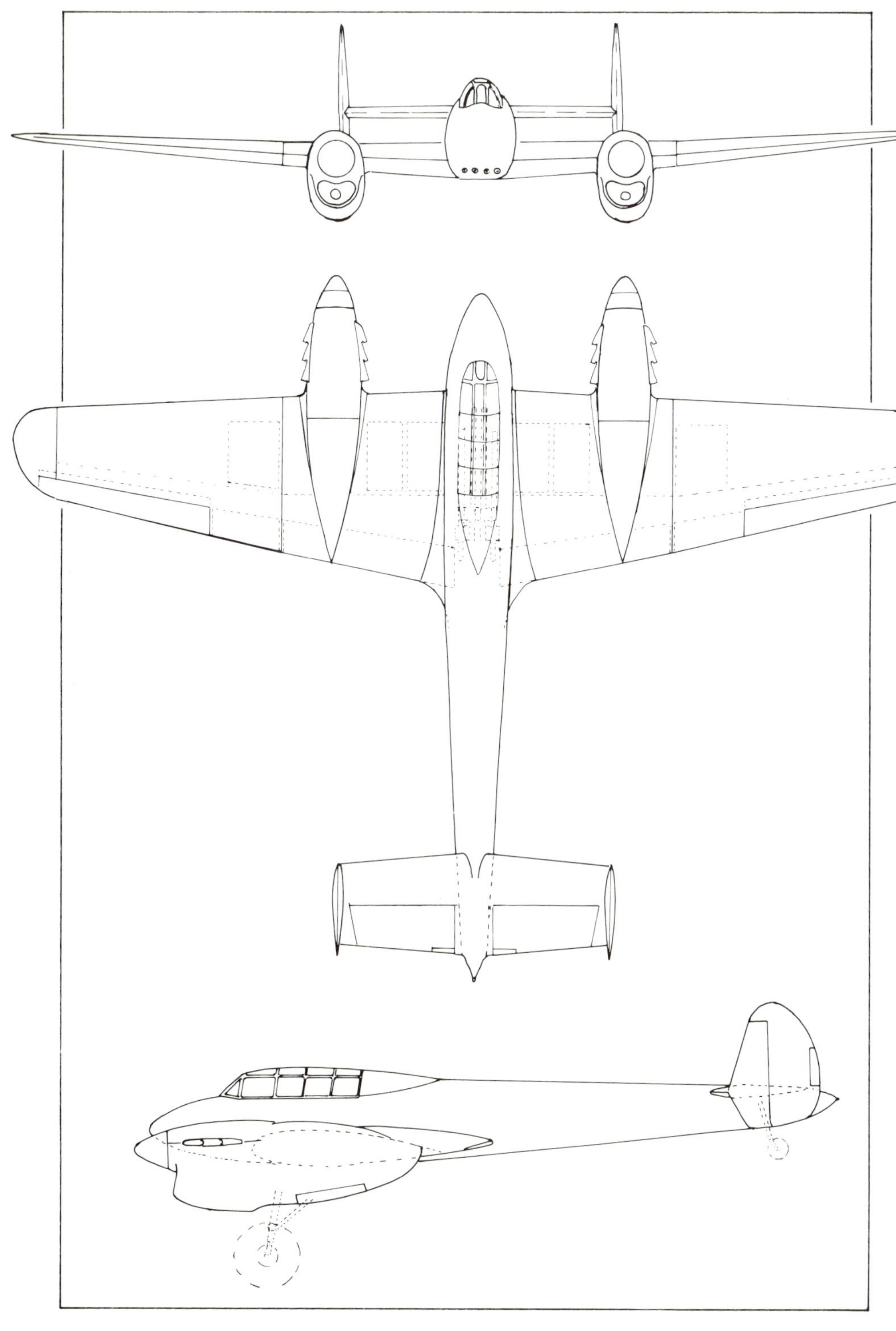

Interceptor fighters for the Royal Air Force

September 23 1940, to invite more tenders and told Gloster that a F.9/37 Merlin-engined and strengthened was worth developing. A two-man crew was now essential, one to handle the radar.

The first Beaufighter to serve in an RAF unit was *R2066* which joined the Tangmere Special Flight in August 12 1940. AFDU Northolt received *R2055* the next day. First squadron deliveries took place on September 2 when *R2072* reached 29 Squadron and *R2073* 604 Squadron. The next day 25 Squadron received *R2057,* but on September 8 600 and 23 Squadrons each took on charge a Beaufighter; the original idea of completely equipping 25 Squadron had changed—the aircraft's problems would be more widely tackled. On October 12, however, it was to 219 Squadron, Redhill, that all 600 Squadron's Beaufighters departed. The first operational patrol by a Beaufighter was flown on September 17 1940 by Squadron Leader S.C. Widdows in *R2072* operating from Digby. Speed trials of production Beaufighters in September 1940 showed them faster than the Hurricane to 3,000 ft and satisfactory to 15,000 ft, where Hercules III examples reached between 312 and 325 mph before their speed fell rapidly away. Also, night bombers had a habit of flying at around 18,000 ft which gave added impetus to desires for an early replacement of the Beaufighter. A decision to rework one F.9/37 was made on October 13, with Boulton Paul aiding Gloster. Six cannon were needed by F.18/40 but only four would fit into the F.9/37, and when its weight exceeded 15,250 lb performance would fast deteriorate. Supplementary .303 in wing guns were a possibility, an observer's dome essential, and the mock-up was officially viewed on October 24.

Two new Beaufighter variants were presently intended, a 1,200-mile range fighter and a long-range extra-tankage version able to fly 1,800 miles and be suitable for Coastal Command whose original requirement had not proceeded. Fighter Command retained priority on the first 100 Beaufighters, the position being reviewed after the 50th delivery. Two Fighter Command squadrons, 25 and 604, would be completely equipped, then one Coastal squadron. *R2153* was delivered to 235 Squadron on November 21, the first to the Command, followed two days later by *R2143.* Both had the Hercules X engines which were fitted to *R2139* and subsequent early Beaufighters. Four cannon, drum fed, were fitted in the first 50 Beaufighter Ifs. Later aircraft could carry two .303 in machine-guns in the port wing and four in the starboard making them more useful in a maritime role, but not all aircraft had them. The intention had been that the 51st *et seq* Beaufighters would have a recoil feed system similar to Bristol's original design which had been rejected due to mistrust of its mechanics. The new system, Mk 1 feed, akin to the French Chatellerault type, was subject to lengthy testing which resulted in a type very similar to Bristol's original. Not until the 401st Beaufighter was completed in September 1941 was recoil feed installed.

Whereas six squadrons had operational Beaufighters in October 1940, Whirlwinds equipped only one non-operational squadron. Sir Roy Fedden tried to persuade Westland to re-engine the Whirlwind but they insisted that larger power plants defeated the purpose of their small, compact fighter whose airframe was unable to accept them. Petter visited HQ Fighter Command on

Opposite *The Gloster Reaper (F9/37 development) two-seat fighter. Two Merlin XX engines. Based on a Gloster diagram of August 19 1940. Wing span 50 ft; length, tail up, 37 ft 10 in; height, tail up, 7 ft 9 in.*

October 2 to discuss the whole Whirlwind programme, the aircraft having been placed second to the Lysander as regards production. Fears that Westland could not mass-produce fighters were regarded as confirmed. Carburettor failures, slat sprocket troubles, wing tip and hydraulic problems, Exactor throttle defects, fractured pipes due to bad fitting at the works and cockpit canopy cracking at rivet lines along the top centre line of the hood, all combined to reduce Whirlwind serviceability to around 58 per cent. The AOC 13 Group thought Westland had done little to remove such irritating troubles but the company countered saying they had a test aircraft flying in fully modified state.

The company's review of the entire programme led to them accepting that an engine change was necessary and they wrote to Fighter Command stating '. . .if you come to think a lot of the aircraft there is always the alternative project for fitment of American radial engines'. But Air Chief Marshal Sir Hugh Dowding thought neither highly of the Whirlwind nor of Westland. He stated in a letter that he had purposely placed 263 Squadron well out of the way of the fighting because he 'knew Westlands' and 'knew what a packet of trouble the squadron would be in for'. He added, 'I cannot put them anywhere in the South because I cannot carry any passengers in that part of the world. All the squadrons there have got to be fighting fit.'

Petter next visited 263 Squadron at Drem, finding two out of eight aircraft serviceable, four grounded due to engine malfunctions and one damaged in a dive. The problem was cured soon afterwards by the fitting of a small acorn fairing at the fin and tailplane junction. Fall-off in performance at height was attributed by Petter to a decline in engine boost, twice that forecast.

Of the 15 Whirlwinds delivered by the first week of November, ten were in squadron hands and Rolls-Royce used one for Peregrine refinement. After the 19th aircraft, production was speeded and the cockpit layout re-arranged to provide for TR1133 and R3003 radio.

Defending their product, Westland told Air Marshal Sir W. Sholto Douglas soon after his appointment as AOC-in-C Fighter Command that theirs was the most radical new aircraft that had ever entered service. Petter wrote '. . .after talking to the Engineering Officers at Drem I really do think that the troubles there have been no worse than, say, with the first Spitfires. But of course, under wartime conditions it is realised they may be far more serious in their effect. Secondly, the Whirlwinds have been moved four times in their short life. The last two stations have been more than 400 miles from Yeovil. This has made our task more difficult and I do hope it will be found possible to allow the very enthusiastic present officers to continue to use the Whirlwind, preferably further south where we can keep in close touch with them.'

Sholto Douglas, not entirely unsympathetic, answered 'The fact remains that it is now five months since 263 Squadron re-formed, allegedly, on Whirlwinds. It still has only 12 and is not yet operational. You must admit this is a sorry story. In order to help you I am moving 263 Squadron, taking its Hurricanes away and making it operational on Whirlwinds. It is up to you to get sufficient aircraft off the production line to bring the squadron's IE strength to 16 at once.' Sholto Douglas, wanting faster production, was alarmed about substantial changes from the 19th machine onwards, which was causing more delay. He told Westland 'The reputation of your firm largely hangs on getting 16 IE aircraft all in working order'. Westland promised the 16 machines by December 14, so on November 28 ten Whirlwinds of 263 Squadron were moved to Exeter and the squadron was declared operational on December 7 1940. At 11:55 hr Red Section comprising three aircraft — *P6974* (Squadron Leader Eeles), *P6976* and *P6075* patrolled the Plymouth area. It had taken four years, ten months to get the first cannon fighter operational.

In contrast, by December 7, 100 Beaufighters had been built, three squadrons were fully equipped and 41 aircraft were in squadron hands. Fighter Command wanted them all fitted with hydromatic propellers whilst many were grounded undergoing a wide assortment of modifications since serviceability rates remained low. *R2054* was the first to fly incorporating Modification 67, the thinner cockpit frame members to improve night vision. Coastal Command's first long-range Beaufighter, *R2152*, became available for trials in late November and the first production Beaufighter fully fitted with fixed Coastal Command fittings was *R2269*, the 217th aircraft, which was delivered in February 1941. About half the Mk Is went to Coastal Command. The Command's initial Beaufighter requests included a rear defence gun, rejected through fear that introducing this, or nacelle-mounted guns, would slow production. Bristol were at the time working on a slimmer fuselage Beaufighter, estimated top speed 357 mph at 20,000 ft.

Increasingly, Coastal Command was involved in shipping strikes and wanted as many cannon-armed aircraft as it could acquire. On December 17 1940 Air Marshal Sir Frederick Bowhill requested a twin 20 mm cannon replacement for four .303 in belly guns fitted to his Blenheims. They could easily be mounted and would be very effective; but reloading in the air was impossible and they would be susceptible to icing, he was told. On January 6 1941 Coastal Command requested a trial installation be made and enquired about belt feeding and electric heating. Boscombe Down built a prototype which 500 Squadron tried. Pressure was put upon the Under Secretary of State for Air on February 21 to enlist his support for the production of cannon fighters to patrol the north-west approaches. Regretfully, the Command was told, on March 26, that surplus cannon would not be available until at least July 1941. Cannon might be taken from Army co-operation squadrons and fitted to 16 Group's Hudsons, but this never came about. Instead, Blenheims had an extra Vickers GO gun, or even two, in the nose, giving a possible six-gun fire system for use during frontal attack, a modification approved on February 27 1941.

In April 1941 No 272, the first Coastal Command squadron equipped with fully modified Beaufighters for maritime use and having Marconi radio, new instruments and a chart table, became effective; followed by No 252 in June, No 248 in July, No 236 in November and No 235 in December 1941. This Beaufighter Ic was about 15 mph faster than the Mk If.

By February 1941 Beaufighter II production was running six weeks late and undercarriage doors, a nuisance on the Mk I, were causing problems. *R2270*, the first production Mk IIf, flew on March 22 1941. A dozen had been delivered by June, by which time 600 Squadron was equipping. Merlin XXs raised the Beaufighter's ceiling by 5,000 ft but the prototypes' speeds were not attained by production aircraft whose top speed was about 332 mph at 30,000 ft. Flame damping was ever a problem on these aircraft, entering service virtually a year later than intended. By December 1941 the production balance of Beaufighters was 38 Fighter Command aircraft to 43 for Coastal Command.

Steady progress was, by 1941, being made with cannon-armed single-seater fighters. Four cannon, drum fed, were ordered in November 1940 for 30

To insure against engine shortage and improve the Beaufighter's performance, the Mk II had Rolls-Royce Merlins. T3144 illustrated was used by Rotol, Staverton, for propeller tests from December 1941 to July 1942, retaining AI aerials. Cockpit framing is thinner than on early Mk Is (Rotol).

Hurricanes to the standard first flown in *V7360* on December 5 1940. Instructions for 32 of the first 100 Yeadon-built Tornados to have cannon were given on December 6, extent limited only by production rates. A hundred sets of four-cannon wings for a new Hurricane, the IIc, were authorised on January 20 1941 and for another 200 in March. The prototype IIc, *V2461*, flew on February 6, its cannon belt-fed.

All cannon Tornados were cancelled on February 17 1941, a simultaneous decision calling for all Typhoons to have them. Supply was limited and the Air Staff order of April 8 1941 that all would mount six cannon caused dismay, and disruption to the whole production programme. It also resulted in production of 12-gun Typhoon Ias whilst the whole production schedule was revised, in time for the Air Staff to settle for four-cannon Typhoons. As with 12 machine-guns, six cannon served only to complicate installation, maintenance and re-arming. The weight of fire from four cannon was adequate.

Beaufighter replacement was subjected to detailed analysis in December 1940. Although the Merlin-powered Gloster F.9/37 was the Air Ministry's favourite, the conclusions on December 18 were that its development would be lengthy and that it would be outdated when available in late 1942. Funding was available for only one design, too. Sir Henry Tizard still favoured the 14,500 lb developed F.9/37 and a new specification, F.29/40, was devised to cover it, outlining a two-seat night fighter whose top speed would be 340 mph at 20,000 ft, its service ceiling 33,000 ft. Armament could be five 20 mm cannon, or two 20 mm cannon and a four-gun BP turret. Wing guns remained optional.

The need for long-range fighters was very apparent during the 1940 Norwegian and French campaigns and now for other overseas theatres too. Accordingly, de Havilland were told on July 18 to complete one of the 50 Mosquitoes, ordered the previous week, as a four-cannon four-machine-gun fighter. F.18/40 prescribed a very similar machine, but to equate F.29/40 meant fitting a dorsal turret. De Havilland were asked if the Mosquito could be thus equipped but, not intending a drag inducer and heavy, unnecessary item to spoil their aeroplane's performance, they stated on December 2 that a turret could not be added without redesigning the aircraft as a low wing fighter.

A turret could, however, be fitted into the Beaufighter, just, but not without performance loss. Pressure mounted to develop that existing type by fitting Griffons in 1942-built examples, giving them '30 per cent extra power'. An alternative was a lightened Beaufighter, still weighing 18,500 lb. A further feasible night fighter was the Griffon-engined Fairey F.18/40 Firefly development of 12,800 lb loaded weight. A top speed of 373 mph at 20,000 ft was claimed. The Air Staff rejected it for the greater night safety afforded by a twin-engined aircraft. In the summer of 1944 the RAF did, incidentally, use a few conventional Fireflies during anti-diver operations.

At Hawker's Langley airfield on December 29 1940 a demonstration was given by new aircraft types, among them the Mosquito and F.9/37. So impressive was the Mosquito's display that early next morning de Havilland were told to complete over half their contract as long-range fighters even though the prototype fighter could not fly before May 1941. During the first month of flight testing the prototype Mosquito had reached 392 mph at 22,000 ft, 'quite phenomenal' commented a privately told guest. A fully loaded fighter was forecast to attain 378 mph at 22,000 ft. The Mosquito fighter was not, at this time, specifically for night fighting.

Another decision of December 30 was that a turretted Beaufighter powered by Griffons or Hercules VIIIs was the best night fighter option. Intention to employ the Martin Maryland for long range escort patrol had vanished at the sight of the Mosquito. Lengthy disbelief in de Havilland's performance estimates meant that it was almost mid-1942 before a Mosquito fighter became operational, initiating a new class of long-range fighter.

In keeping with December's decision a Griffon Beaufighter was tested, but the superiority of the Mosquito caused cancellation of that programme. The turret fighter lobby had its way too and in April 1941 two Beaufighter IIs and two Mosquitoes were ordered, each carrying a BP turret immediately aft of the cockpit. Prior to modification, Beaufighter *R2274* had a top speed of 335 mph at 21,000 ft. With the four-gun turret in place and only two cannon remaining, *R2274* had a flying weight of 18,695 lb and a maximum speed of 302 mph at 19,300 ft. With the Mosquito doing very well and ample Beaufighters to hand, the Gloster F.29/40 Reaper programme was halted on May 1 1941, along with major new Beaufighter variants for night fighting.

Of the numerous Beaufighter refinements made in 1941 the most obvious was the 12½° dihedral tailplane. Powerful engines and inherent aerodynamic features induced longitudinal instability, particularly in the tail-heavy Mk II. A twin-tail arrangement was tried on *R2268*, but the dihedral tail plane tested on *R2057* (Hercules II) was chosen ensuring good stability to 140 mph IAS. This was not universally popular in Fighter Command since it restrained manoeuvrability, but it proved very popular with Coastal Command crews.

At 14,400 ft, full throttle height of Hercules III-engined production Beaufighters, their top speed was about 323 mph, service ceiling 27,900 ft. Mk Ifs fitted with Hercules Xs and XIs using 100 octane fuel weighed (fully loaded) 19,550 lb, their peak speed being 335 mph at around 15,500 ft. These figures related to four-cannon Mk Ifs, and adding six guns reduced their speed to around 320 mph.

The first Hercules VI Beaufighter trial installation was underway in February

Above *Night fighter Beaufighter Ifs gave better service than expected, the capacious body being ideal for early, bulky AI radar. X7583 served with 68 Squadron as WM-E from August 1941 until March 1943 when it joined 51 OTU and stayed in use until February 1945* (RAF Museum P012519). **Below right** *Beaufighter If X7579 carries centimetric radar, AI VII, in its thimble nose radome. No Yagi aerials were then needed. The aircraft served with TFU, FIU, 29 Squadron from July 1942 to March 1943 and later at 51 OTU* (BAe).

1941. Numerous problems delayed it and not until October 5 1941 did the aircraft, *X7542*, proceed to A&AEE for acceptance trials. Early hopes for greatly improved performance were not realised. *X7542* (mean flying weight 19,750 lb) had a top speed of 333.5 mph at 15,600 ft and reached 326.5 mph at 22,000 ft. It took 7.5 mins to climb to 15,000 ft, 25,000 ft being reached in 19 mins. Production standards for the Beaufighter VIf/P2 were agreed on April 9 1942 and, soon after, the first Mk VIs joined Fighter Command. Mk II production ended in April 1942 and the Mk I in July 1942. By then Beaufighters were proving most useful to Coastal Command and valuable in the Mediterranean theatre too. Their future lay increasingly in the strike fighter role and when sufficient Mosquito night fighters were available the Beaufighter VI was replaced in Fighter Command, the last squadron, No 409, re-equipping in June 1944. In retrospect the cannon fighter 'Beau', which served in the Far East until September 1945, can be seen to have been a most useful carrier of bulky AI radar and later of centimetric Mk VIII AI situated behind a 'bull nose' type radome.

What of the prime cannon fighter, the Whirlwind? As 1940 was ending Westland at last agreed to fitting Merlin Mk XXIIs to a six-gun version, forecasting 410 mph, 37,000 ft ceiling and 800-mile range. Fighter Command interest acknowledged the Whirlwind's failure as attributable to under-developed powerplants. The aircraft's service record was scrutinised and Sir Hugh Dowding, knowing the Whirlwind's performance above 25,000 ft was poor, asked 263 Squadron's commander for his views. He confirmed that performance fell away above 20,000 ft, rapidly after 26,000 ft, although one

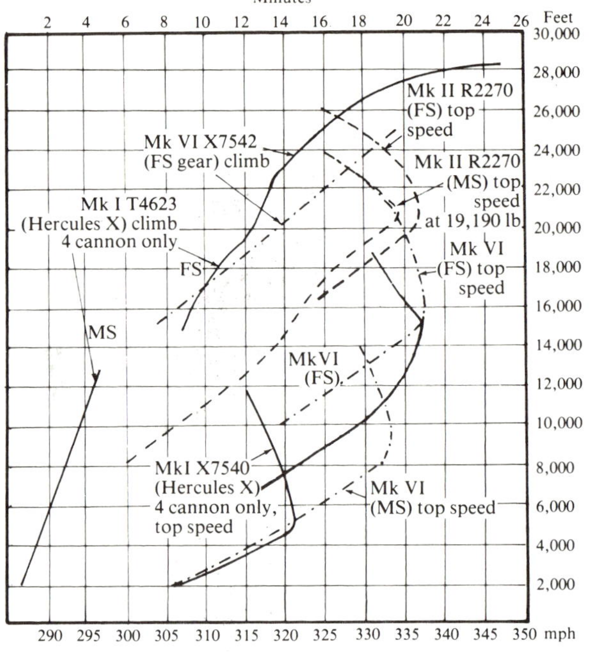

Graph of the Beaufighter's performance, comparing the IF, IIF and VIF versions.

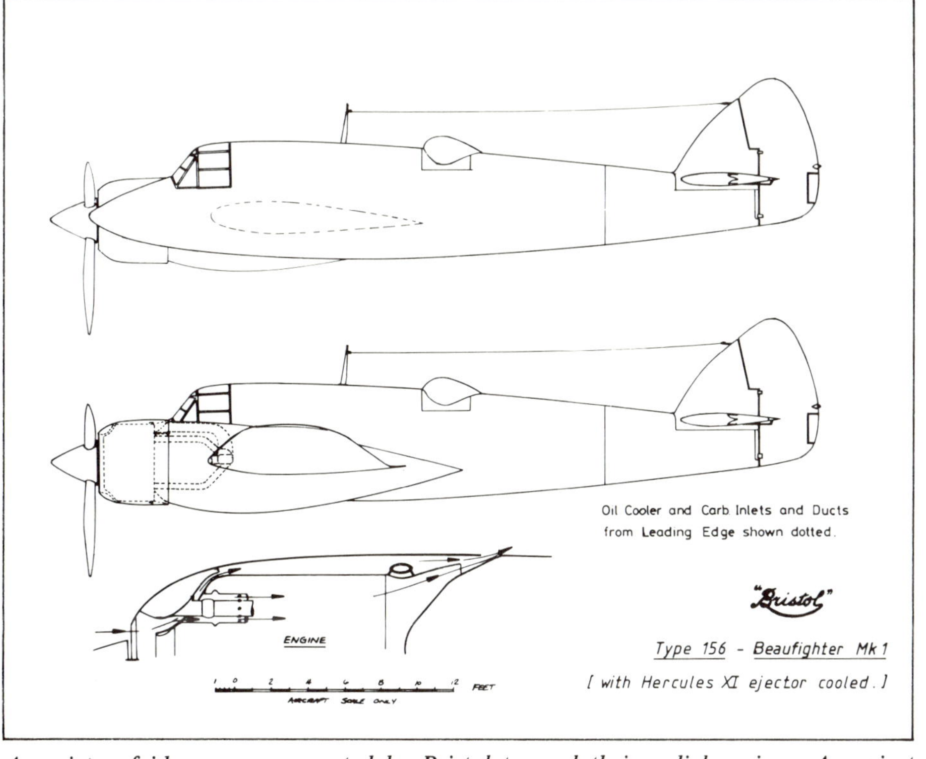

A variety of ideas were promoted by Bristol to cool their radial engines. A project illustrated here was for ejector cooling applied to a Hercules XI-powered Beaufighter Mk I (D.C. Greenman).

pilot had managed 27,000 ft before engine problems occurred. Concern at tail vibration in dives led to testing of *P6980* flying at 10,144 lb after which all Whirlwinds had an enlarged acorn at the fin and tailplane junction, a feature of *L6845* and the seventh and subsequent aircraft.

Whether to persevere with a few Whirlwinds was also the concern of Air Marshal C. Quinton Brand, AOC 10 Group, who asked Squadron Leader Munro to update his opinion. On having the report Brand considered the squadron had too high an opinion of the machine. Its automatic slots, he thought, introduced a false sense of security for their drag disadvantaged the aircraft if trying to out-manoeuvre single-engined fighters. Linking flap and radiator control seemed unsound because a pilot forced to make a second approach was obliged to close the flaps when most needed. On discovering it took 200 hrs to change an outer mainplane and three to four days to make an engine change, Quinton Brand did not hesitate to comment 'I would be sorry to see any extension of the use of the Whirlwind in service'.

Squadron Leader Munro, however, based his assessment on seven months' experience and two years' fore-knowledge. He considered that in experienced hands the Whirlwind was a gentleman's aeroplane. He thought its take-off good and the climb to 10,000 ft exceptional, being reached in about four minutes. At that height it was faster than a Spitfire — much faster when lower.

Manoeuvrability, he said, was better than the Hurricane's or Spitfire's above 360 mph. View from the well laid-out cockpit was very good, and general serviceability not bad. If only the engines had at least given the expected power!

From these mixed opinions Fighter Command had no wish to see production extended but decided to persevere perhaps using the Whirlwind in two squadrons 'outside the radius of action of enemy fighters'. Air Vice-Marshal H.I.T. Beardsworth, CTO, reckoned the four cannon in a 340 mph fighter would be most effective against enemy bombers coming in at around the usual 18,000 ft in resumed day raids. Additional fuel and 120 rpg meant the Whirlwinds could be used at far greater ranges than existing fighters. Investigations suggested a Merlin development would weigh 12,000 lb, its wing loading being 48 lb/sq ft. To accommodate larger, necessary propellers, meant major wing redesign with loadings rising dramatically during take off. Redesign of the wing to give it greater area would improve the ceiling only to reduce the speed, so the proposition was turned down. At a meeting on February 3 1941 the Joint Development and Production Committee, although 'sympathetic to it', decided that Whirlwind production should cease in August. It required 50 per cent more material to perform the same duty, less efficiently, than the Spitfire, and even present production was inadequate to meet wastage.

What of the 114 Whirlwinds to be built? For many months they flew defensive and coastal patrols. May 1941 brought two disturbing slot failures so the A&AEE was ordered to investigate flying characteristics with outboard slots closed. Trouble had come when a slot track sheered. Operated by aerodynamic forces alone, the slots were supported by smaller slots near the engines and connected to Fowler flaps so that when those extended the slots went progressively forward. Now, with the outer slots locked, take-off and landing speeds were not affected. Manoeuvring showed little difference without free slots, so all were fabric sealed to improve stall characteristics. In the vicinity of the aircraft's self-sealing fuel tanks the mainplanes were thickened to accommodate larger tanks, possible effects on flying being now investigated using *P6997*. No firm conclusions were drawn, but testing this production machine showed its top speed as 338 mph at 15,200 ft, lower than the prototype's, probably due to additional equipment and possibly to it being a sub-standard example.

During the latter part of 1941 Whirlwinds used their cannon for attacks on Continental targets, airfields in particular. At the start of January 1942 all fighter squadrons were ordered to train for night operations. Previously the Whirlwind was reckoned unsuitable for night operations but higher landing speeds were now acceptable. On January 29 Squadron Leader Coghlin of the second Whirlwind squadron, No 137, after a night flight, considered the aircraft quite suitable. Whirlwinds were to commence night interdictor-duty on November 21 1942.

For offensive operations Whirlwinds were of limited value because they did not carry bombs. Such additions had been suggested by Squadron Leader Pugh in September 1941 but rejected because fighter-bombers were in their infancy and the 'Hurribomber' had yet to go into action. Its success was such that in June 1942 Westland were asked to investigate the possibility of fitting bombs to Whirlwinds. On July 3 Fighter Command requested that Whirlwinds of both squadrons be fitted with bomb racks. Westland worked fast, stating that modification was possible and parts easy to make so the go-ahead was given at the start of July. Trial installation was completed at Yeovil on August 4 and approval then given to make parts for 50 aircraft. Both 250 lb and 500 lb bombs were to be carried, of the short-tailed variety designed for Mosquitoes. Fitting the equipment took longer than expected, the TI aircraft *P6997* reaching A&AEE on August 15. Handling trials with two 500 lb bombs aboard showed the 'Whirlibomber' safe to fly at up to 350 mph IAS. Above that speed it flew with left wing very low and encountered aileron snatch. Take-off weight was 11,215 lb loaded; time to 12,000 ft 5.4 min; maximum speed (FS) 318 mph at 15,000 ft and service ceiling about 27,500 ft. Take-off run was extended by about 125 yds.

On the afternoon of September 7 1942 No 174 and 175 'Hurribomber' squadrons were released from operations, 263 Squadron taking their places. Next morning three Whirlwinds took off on an anti-shipping *Roadstead*, the first operation during which bombs were carried. Each carried two 250 lb bombs, but their tanker target was already in Cherbourg. Next day four 'Whirlibombers' carried out the first attack on four armed trawlers off Alderney. No 118 Spitfire Squadron acted as anti-flak guard while Squadron Leader Woodward and Flight Lieutenant G.B. Barnes attacked the second of the vessels. Woodward fired all his cannon shells, his bombs falling close to a ship. Two others attacked the leading ship and it later transpired that two trawlers were sunk.

P7012 being loaded here served with 137 Squadron from November 1941 to January 1943, then went to 263 Squadron. It had the distinction of being the last Whirlwind to be recorded as flying an operational sortie, on November 29 1943, flown by Pilot Officer Blacklock during Operation 'Hanwich', against Ju 52 minesweepers reckoned to be operating off Cherbourg (RAF Museum P9516).

Fighter Command confirmed, on September 11, that all Whirlwinds would be converted into fighter-bombers, No 137 Squadron commencing bombing raids on October 12. Rarely did Whirlwinds engage in combat but on January 14 1943 Flying Officer Koyne and Sergeant Cotton were flying anti-*rhubarb* patrols between Portland and Aldhelms Head when they were vectored to 20 miles north of Barfleur. Turning about they met two Fw 190s and a low-level dog fight ensured, the first such battle and the outcome was inconclusive. Not since Pilot Officer King had shot down a Bf 109 a year previously had such action come about, and a reduction in the Whirlwind's radius of action from 165 miles to 135 miles was about to take place.

Squadron Leader Baker of 263 Squadron flying *P7040* led two other Whirlwinds to intercept Ju 52 minesweepers off Cherbourg on November 29 1943. Nothing was seen and as they landed at Warmwell they brought the Whirlwind's operational career to a close, the aircraft being put up for disposal, at Ibsley, on December 19 1943. Between September 1 1942 and November 30 1943 the two Whirlwind squadrons flew 1,427 sorties, 349 at night and 698 against shipping. Flying accidents resulted in 26 aircraft (22.8 per cent) being written off. Ten more were written off due to battle damage and 33 (28.9 per cent) of those built failed to return from operations. Of the total built 39 had been written off by September 1 1942.

An official appraisal of the Whirlwind upon its retirement attributed 50 per cent of its failure to the engines. Petter had claimed that it should have started around Merlins, and with a higher safety factor. Instead, it was too small and weak for development and additional equipment. The Air Ministry gave Westland the impression it was not wanted, particularly by the operations staff. Westland considered the first squadron did not give it a fair chance by failing to work out tactics suitable for a twin-engined day fighter. Mid-war years brought a trend towards twin-engined fighters so that in a number of respects the Whirlwind was ahead of its time.

On January 1 1944 it was declared obsolescent and in June 1944 obsolete, the 16 survivors being in store at 18 MU, Dumfries, where they languished until scrapped late in 1946 — except for *P7048*. That became *G-AGOI*, Westland's 'hack', and received a Certificate of Air-worthiness on October 23 1946 but was dismantled at Yeovil in May 1947. 'The cannon fighter' had passed into history.

In retrospect

Piston-engined interceptor development showed the overwhelming importance of powerful, suitable engines. The superb Rolls-Royce liquid-cooled Merlin made possible high speed monoplane fighters, then the prospect of the Napier Sabre providing double the power made this an even more attractive position. The Sabre's failure dealt a mighty blow at the projected fighter force from which it did well to recover. Hawker's hefty fighters demanded the Sabre's might to shift them, before its failure brought unscheduled reversion to the air-cooled radial which, all along, had its devotees in official places. Although the Bristol Centaurus was a long time in coming it performed well, removing the need for elaborate cooling systems which plagued liquid-cooled engines and made them more vulnerable in combat. Almost unnoticed, aircraft came into being that were bullied along by brute force. High drag, the hallmark of radial engines, was partly overcome by careful cowling design more than by novelty items such as ducted spinners.

Powerful engines were naturally thirsty, aggravating the failing of so many British fighters. Their lack of range was due to their being designed to defend limited air space, mainly by flying costly standing patrols. That situation was brought about by a desire for speed at the expense of rapid climb. Mid-war attempts to devise lighter fighters went some way to improving duration possibilities, but more important was the development of long-range fixed and droppable external tanks. Necessary plumbing was easiest to fit into the thick, drag-inducing wings of Hawker fighters.

Switching to wing guns was, without doubt, a major advance possible only as a result of more reliability in weapons. It only came to be seriously questioned when the first jets with closely grouped nose guns—Mosquito and Whirlwind fashion—showed greater accuracy, because their fire was easier to align. The Hurricane, with wing guns grouped, had an advantage over the Spitfire which was retained when increased calibre guns could be more easily fitted into its wings. Too many guns led to wasteful practice, serious production and design problems and higher levels of unserviceability, eight machine-guns or four cannon proving sufficient. Some benefit might have arisen through switching sooner to .50 in guns in favour of .303 in, although any weight increase would have reduced the fighter's performance, maybe withdrawing the slight advantage.

Unforeseen until 1940 was the ability of fighters to replace light bombers, German Bf 109s demonstrating the possibilities. Hawker's heavy fighters again had an edge over the nimble Spitfires, being easily adaptable for the ground-attack and support duties which were to become so important. The Tempest II would beyond doubt have excelled in that role, whereas the Spitfire's speed suited it more to fast tactical reconnaissance than bombing. In the Far East Hurricanes gave good account of themselves, along with Beaufighters whose strike role was particularly useful at sea.

No consideration of RAF interceptor fighters should fail to amply acknowledge the Spitfire's part in the story. Here was an aeroplane whose all-up weight, incredibly, doubled; whose speed increased by 100 mph and whose interceptor qualities were superlative. It was developed with consumate skill over six years more than the Hurricane. Not only did it intercept the enemy but it played an equally important part protecting the huge fleets of British and American bombers that pounded the Continent in daylight during 1943-44, before the fighter-bomber largely replaced the light bomber. Perhaps even more important was the boost to morale that the Spitfire's beautiful form and unmistakable whistle brought to the darkest weeks of the war. That was something the numerically greater Hurricane never achieved, and it was 'Spitfire' that became almost synonymous with the 'Battle of Britain'.

Sad indeed that, when the fighting ended, Supermarine's future was so bleak with only the troublesome Spiteful and a jet derivative on offer. Never again was Supermarine to build an aeroplane of such excellence as the Spitfire. When the RAF chose its main postwar fighter it purchased it from Hawker. Supermarine was merely asked to provide a stop gap until the Hunter was ready.

Appendix

Official statistics relating to fighter delivery and service use

Interceptor fighters and their derivatives in front-line home-based squadrons, 1940-44

In the details given below the first item denotes the number of squadrons, the second the number of aircraft equipping them.

	1.4.40	1.9.40	1.1.41	1.4.41	1.7.41	1.10.41	1.1.42
Gladiator	2/66	1/11	1/10	—	—	—	—
Hurricane	24/414	37/703	40/754	40/609	30/623	32/616	12/265
Spitfire*	19/322	19/349	19/347	24/484	37/722	43/959	58/1,141
Whirlwind	—	−/5	1/14	1/20	1/21	2/32	2/42
Beaufighter	—	—	/48	4/77	6/114	6/10	NA
Typhoon	—	—	—	—	—	/4	1/14

	1.4.42	1.7.42	1.10.42	1.1.43	1.4.43	1.7.43	1.10.43
Hurricane	12½/187	12½/269	15/199'	12/202	4½/51	3½/52	4½/92
Spitfire*	58/1,189	59/1,203	58/1,093	47/895	48/879	48/902	48/963
Whirlwind	2/38	2/38	2/41	2/35	2/36	1/19	1/19
Beaufighter	NA	NA	18/353	14/258	11/228	8/155	7/126
Typhoon	2/29	3/62	8/117	13/231	16/352	19/350	18/295

	1.1.44	1.4.44	1.7.44'	1.10.44'
Hurricane	5½/84	3½/59	1½/30	1/20
Spitfire*	51/939	54/1,018	63/1,216	57/1,133
Whirlwind	1/16	—	—	—
Beaufighter	5/102	3/57	2/25	—
Typhoon	20/377	22/376	20/362	21/401
Tempest V	—	−/13	2/50	6/119

Notes:
*Includes all Spitfire variants including Mk VI/VII where relevant.
†Includes aircraft operating in France with 2 TAF.
The table excludes long-range fighters, turret fighters and Army co-operation support fighters. Details relating to Mustang III/IV aircraft NA (not available). On June 6 1944 2 TAF held 594 fighter-bombers and 324 aircraft equipped to fire rocket projectiles. The strength of the RAF's interceptor fighter force on June 30 1938 makes interesting comparison. The three Fury squadrons held 42 aircraft, nine Gauntlet squadrons had 126 aircraft, seven Gladiator squadrons held 79 aircraft and the three Hurricane squadrons 41.

Delivery of new fighter aircraft in the United Kingdom

	Hurricanes	Spitfires	Typhoons	Tempests	Beaufighters
1939	569	434	—	—	—
1940	2,521	1,246	—	—	111
1941	3,167	2,518	28	—	800
1942	3,067	4,134	686	—	1,579
1943	2,742	4,276	1,137	3	1,641
1944	688	4,916	1,165	504	1,155
1945	—	2,627	299	702	271

These figures relate to the numbers of fighter airframes delivered complete to the RAF in Britain and exclude airframes packed immediately for overseas delivery. Beaufighter listing includes aircraft for all Commands in the United Kingdom.

Total stock of fighter aircraft held at home and overseas on May 31 1945

	Strength*	In MAP hands/in transit between theatres	Reserves, unavailable in less than 60 days	Total
Beaufighter'	1,201	73	146	1,420
Gladiator (for met duty)	7	—	—	7
Hurricane	1,565	33	186	1,784
Kittyhawk	153	1	96	250
Meteor	58	12	—	70
Mosquito†	2,757	490	161	3,408
Mustang	876	293	86	1,255
Spitfire	5,015	339	510	5,864
Tempest II/V	423	67	48	538
Thunderbolt‡	555	44	39	638
Typhoon	1,107	42	17	1,166

Notes:
*Includes aircraft which can become available in less than 60 days.
†Includes all versions in all Commands.
‡Included, although used mainly as strike aircraft.

Index